Industrial Mathematics

AUSTRALIAN MATHEMATICAL SOCIETY LECTURE SERIES

Editor-in-chief: Associate Professor Michael Murray, University of Adelaide

Editors:

Professor C.C. Heyde, School of Mathematical Sciences,
Australian National University, Canberra, ACT 0200, Australia

Associate Professor W.D. Neumann, Department of Mathematics,
University of Melbourne, Parkville, Victoria 3052, Australia

Associate Professor C.E.M Pearce, Department of Applied Mathematics,
University of Adelaide, SA 5005, Australia

Industrial Mathematics

Case Studies in the Diffusion of Heat and Matter

GLENN R. FULFORD

PHILIP BROADBRIDGE

CAMBRIDGE
UNIVERSITY PRESS

CAMBRIDGE UNIVERSITY PRESS
Cambridge, New York, Melbourne, Madrid, Cape Town, Singapore,
São Paulo, Delhi, Dubai, Tokyo, Mexico City

Cambridge University Press
The Edinburgh Building, Cambridge CB2 8RU, UK

Published in the United States of America by Cambridge University Press, New York

www.cambridge.org
Information on this title: www.cambridge.org/9780521001816

First published 2002

A catalogue record for this publication is available from the British Library

Library of Congress Cataloguing in Publication data
Fulford, Glenn.
Industrial mathematics : case studies in the diffusion of heat and matter / Glenn R.
Fulford, Philip Broadbridge.
p.   cm. – (Australian Mathematical Society lecture series ; 16)
Includes bibliographical references and index.
ISBN 0-521-80717-4 – ISBN 0-521-00181-1 (pbk.)
1. Engineering mathematics–Industrial applications.
2. Heat–Transmission–Mathematical models.   3. Mass transfer–Mathematical models.
I. Broadbridge, Philip, 1954–   II. Title.   III. Series.
TA331 .F85   2001

ISBN 978-0-521-80717-3 Hardback
ISBN 978-0-521-00181-6 Paperback

"Dedicated to my niece, Elise Campion"

G.F.

"Dedicated to Matthew and Daniel, the two fine sons of Phil and Alice"

P.B.

Contents

Contents

Preface

At a pragmatic level, there are often no alternatives to mathematical models to test new industrial designs. Physical prototypes may be too expensive or too time consuming. In both the design phase and the operations phase, the direct measurement of important operational variables may be impossible or uneconomic. The regions of interest may be inaccessible because of mechanical barriers, high temperatures or hazardous chemical environments.

Mathematical modelling is an efficient and relatively inexpensive device for testing the effect of changing operating conditions in an industrial process. It is far easier and less costly to change a small number of parameters in a mathematical model than to shut down an industrial plant and modify its large-scale equipment. In most circumstances this should not be done as a trial-and-error experiment.

In this era of rapidly changing technology, the efficacy of mathematical modelling in industry should be appreciated more than ever before. Therefore, it is frustrating to note a recent trend towards reducing the amount of core mathematics subjects in engineering degree programmes. It is our expectation that this book could at least be used for an optional course for the more mathematically-oriented students of engineering. Such a course is even more important in the education of those with an interest in the logical design of new technology when the majority of graduates have insufficient mathematical training for this purpose. This course would meet one of the needs of a modern industrial mathematics course, namely that of relevance to real industrial problems. This would alleviate a common criticism by recent engineering students that little attempt is made to relate mathematics service courses to their professional practice.

On the other hand, real-world mathematical modelling courses are good preparation for mathematics students who hope to work in the industrial environment. In too many mathematics degree programmes, applied mathematics courses are presented solely as mathematical methods courses, i.e. solution techniques for various types of equations and optimisation problems. In typical methods courses, artificial applications are often added almost as an afterthought following long sessions of formal theoretical development. Like most mathematicians, we can understand the power and beauty of mathematical rigour in the development of useful mathematical methods. However, without some practice in real-world mathematical modelling, mathematics graduates do not have the best preparation for work in industry.

In the industrial context, mathematicians are not presented with a set of equations ready to be analysed. Instead they are confronted with a set of practical problems that have not yet been expressed in mathematical terms. This translation to mathematical terms is a difficult step if one has had no experience at this activity. From our dealings with many scientists, engineers and mathematicians, we have found that the most proficient mathematical modellers are the experienced modellers, who have learnt to listen to specialists from other non-mathematical backgrounds. However, we still believe that good mathematical modelling skills are based on fundamental principles that can be taught in a course based on case studies.

Good mathematical models must respect accepted scientific laws such as physical conservation laws. Insight can be gained by idealising a model so that only the most important factors, processes and parameters are retained. This is also an aim of experimental control. Secondary effects may be added later as perturbations on the leading terms. Mathematical predictions of physical processes (and perhaps of economic and behavioural activity) should ultimately be expressible in terms of dimensionless variables and the key factors must be expressible in terms of dimensionless parameters.

This text represents a course that has undergone many modifications since 1986 after the authors have presented it to third year mathematics students at three Australian universities; La Trobe University, The Australian Defence Force Academy (University College of The University of New South Wales) and The University of Wollongong.

We will comment on some features of the course. First, industrial case studies are presented at the outset rather than as add-on examples. These case studies lead to mathematical models that motivate development and reinforcement of mathematical methods. It is our experience that the students understand the mathematical methods better after they have applied them to case studies. The level of mathematics used here is not advanced; some of it will already have been encountered at second year level. Assumed mathematical techniques include exact solution methods for constant-coefficient ordinary differential equations, systems of linear algebraic equations, graphical solution of nonlinear transcendental equations.

Other mathematical techniques, that many students may not have seen at second year level, are introduced in a rudimentary way. These include:

- free boundary value problems for partial differential equations (Chapter 2);
- Stretching transformations (Chapter 3);
- perturbation expansions (Chapter 4);
- bifurcation analysis (Chapter 5);
- Fourier series and nonlinear transformations of nonlinear partial differential equations (Chapter 6).

These chapters, containing the case studies, are self-contained and may be studied in any order (after Chapter 1) to suit the backgrounds of the students. However, the techniques introduced in Chapter 3 are an extension of those introduced in Chapter 2, so Chapter 3 should ideally be studied after Chapter 2.

Given the necessity of placing no more than reasonable demands on students, it has been possible to concentrate on only one illustrative area of activity in industrial modelling. We have chosen this area to be continuum modelling involving diffusion and heat conduction. Of course it would be equally possible to choose another area such as queueing theory, operations research, number theory or coding theory. In order to make our course accessible to mathematics students with very little training in physics, we have had to make brief oversimplified accounts of the underlying physical processes. Also, we have not done justice to the techniques of numerical analysis that are useful tools in mathematical modelling. Almost all of our students have recently or concurrently taken courses in numerical methods and we have referenced these in our lecture classes.

As evidence that this kind of applications-oriented course is uncommon, many of our students have stated that this course is unlike any other. However, after taking this course, many have felt a greater degree of confidence in their mathematical skills and in applying them to industrial problems.

We wish to thank our colleagues who have given useful suggestions for the development and improvement of this course. Among them are Yvonne Stokes, David Clements, Kerry Landman, Stephen Bedding, Timothy Marchant, Edgar Smith, Rodney Weber and Havinder Sidhu. We would also like to thank Bradley Loh and Lance Miller for their assistance with proof reading the manuscript. We would also like to thank our editor, Roger Astley.

We have presented this course to several hundred students. Many of these have inspired improvements through their feedback. Their professional development has been a source of great pleasure to us.

Glenn Fulford,   Philip Broadbridge,
University College, ADFA.   University of Wollongong.

1

Preliminaries

In this chapter we set the scene by introducing the case studies of the following chapters. We also introduce the main physical concepts for diffusion and heat conduction, and show how to formulate the main partial differential equations that describe these physical processes. Finally, dimensionless variables are introduced and it is shown how to scale differential equations and boundary conditions to make them dimensionless.

1.1 Heat and diffusion — A bird's eye view

Here we give a basic physical description of mass transport and heat transport by diffusion. This provides the physical ideas needed to formulate an appropriate differential equation, which is done in the next chapter.

Diffusion

Diffusion is a physical phenomenon involving the mixing of two different substances. Some examples include salt in water, carbon in steel and pollution in the atmosphere.

A fundamental quantity is the **concentration** of one substance in another. This may be defined in several different ways. For example, the concentration can be measured as the ratio of the mass of one constituent to the total volume of the mixture (kilograms per litre). Another measure of concentration is the volume of one constituent to the total volume of the mixture.

Due to the random motion of constituent particles, concentrations tend to even out. Some molecules in a region of higher concentration move into a region of lower concentration. (See Figure 1.1.1).

particles diffusing

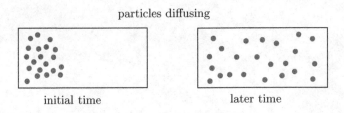

initial time   later time

Fig. 1.1.1. The mechanism of diffusion — due to random motion of particles a high concentration redistributes towards a region of lower concentration.

Heat and temperature

An important thing to remember about modelling heat transport is that heat and temperature are **not** the same thing. Heat is a form of energy and may be measured in joules (the SI unit of energy). The heat energy of a rigid body is the kinetic energy due to the internal random motion of many vibrating constituent molecules. As heat is added to the body, energetic molecular collisions occur more frequently.

In the kinetic theory, temperature is interpreted as a measure of the average internal kinetic energy of constituent particles. The total heat energy is proportional to the temperature of an object and its mass; the latter being a measure of the number of particles. Temperature is a property that determines the *rate* at which heat is transferred to or from the object. Heat energy flows from hot (high temperature) to cold (low temperature). The temperature is defined according to a scale which depends on the expansion properties of certain materials. Temperature is usually measured in degrees Celsius ($^\circ$C) or kelvin (K). Thus $10\,^\circ$C means that mercury in a thermometer will rise to a given height, representing this temperature. Note that $0\,\mathrm{K} = -273\,^\circ$C. The Kelvin scale is designed to mean that $0\,\mathrm{K}$ corresponds to zero internal vibration (absolute zero).

This bird's eye view has deliberately been sketchy and incomplete. For more information on the kinetic theory of gases, the interested reader may consult almost any general introductory physics texts, such as Halliday and Resnick (1974). For the more general theory of thermodynamics, see for example (Feynman et al., 1977, Chapters 42–44).

1.2 Mathematics in industry

In this section we will briefly discuss general opportunities for applied mathematics in industry before focusing specifically on mathematical problems in heat and mass transport in the next section.

Opportunities for mathematicians

Mathematics is a subject that has been studied for several hundreds of years. Much new mathematics has been motivated by practical problems. On the other hand, mathematical models have also been used by industry to improve production, increase profits and generally improve understanding of complicated processes. There is a clear benefit to both mathematics and industry arising from the application of mathematics to industry.

In some countries (for example, Australia and New Zealand) industry puts less effort into research and development than do most other industrialised nations. However, this deficiency is now widely recognised in those countries and some remedial steps have been taken. Recent governments have provided various taxation incentives and assistance schemes for private companies to invest in their own research and development (although, more recently, this has unfortunately been cut back). This has opened up more employment opportunities for scientists, including applied mathematicians. Universities have made efforts to improve their level of collaboration with industry by setting up Industry Liaison Committees and forming consulting companies. Another source of contact between industry and academia throughout the world occurs through Mathematics and Industry Study Groups, pioneered at Oxford University in the United Kingdom. These bring together academics and representatives from industry to apply mathematics to industrially important problems in problem-solving workshops.

There are great benefits to be gained from employing applied mathematicians in industry. Optimisation skills are particularly important on the financial side. For engineering engineering many technical problems can be formulated as mathematical problems and thereby analysed and solved more efficiently. Mathematical models can be used to help understand the underlying physics, chemistry and biology of some processes. This understanding can then help to make the process more efficient. The financial savings can be considerable.

The applied mathematician working in an engineering or scientific environment must be a 'Jack' (or 'Jill') of all trades. That is, she or he must have good scientific general knowledge and also be skilled at formulating mathematical descriptions of practical problems. One advantage that an applied mathematician has is that because mathematics is a universal language he or she is able to communicate with other scientists from a wide variety of disciplines. The applied mathematician must be willing to be guided by other scientists in a team as to which physical variables are important and which directions the research should take once the initial mathematical model has been set up and validated.

Traditionally, applied mathematics students are taught mathematical methods and these are practised on standard problems which are already posed in mathematical form. It is more difficult to train someone to carry out the important first step of mathematical modelling, which is to take a practical problem and simplify or express it in a form which is amenable to mathematical analysis. Proficiency in formulating problems is usually obtained only after years of practice. However, there are some general principles which can be applied to some broad classes of problems, and these may be learnt. For example, heat and mass transport is based on the principle of conservation of energy and matter. As may be seen from reports of industrial study groups, there is considerable demand in industry for skills in this area.

1.3 Overview of the case studies

In this book we will restrict ourselves to modelling those processes which involve transport of heat energy or mass. Industry provides many examples of the use of the standard equations of heat and mass transport and sometimes suggests interesting modifications to the basic theory.

Our primary aim is to study the industrial case studies that are described below. We will along the way, however, consider various other simpler industrial problems, as we develop sufficient physical and mathematical expertise with the phenomena of heat and mass transport. After developing skills for formulating appropriate partial differential equations we consider some analytical techniques for solving them.

Analytical techniques are useful for gaining physical insight. For very complex problems, numerical approaches are often used. It is often useful to start with a very simple model of a complex system whose equations yield an analytic solution. Then a more realistic model can be solved numerically. Together with the analytical results for the simpler models, the numerical results can yield maximum insight into the problem.

Continuous casting

One of the cases that we will study (Chapter 2) concerns a proposed technique of casting steel by pouring molten steel onto a cooled rotating drum. This is done to produce sheets of steel that are longer (and thinner) than those produced by pouring molten steel into moulds. The question we will try to answer is — under what circumstances will the process work? We will do this by predicting how fast the molten steel solidifies.

Water filtration

One method of extracting salt from water is to use a process called reverse osmosis. This involves water passing through a semi-permeable membrane and leaving the salt behind. In this process a major problem is that the salt accumulates at the semi-permeable membrane and restricts the passage of water through it. We will develop a simple diffusion model (Chapter 3) in an attempt to predict the salt buildup along the semi-permeable membrane. To do this we will introduce the method of stretching transformations as a method for solving the resulting diffusion equation.

Laser drilling

Another major case study we consider (Chapter 4) is where a high intensity laser or electron beam is focused on a sheet of metal. The laser drills a hole through the metal and we wish to predict how fast this occurs. This problem is of great interest in many industries where lasers are now being used for cutting and welding.

Factory fires

In another case study we will look at the previously unexplained sudden onset of fires in a New Zealand chipboard factory (Chapter 5) . The aim here is to determine if ignition can occur due to the heating of dust piling up on hot presses. Oxidation of the dust creates heat which may cause the dust to ignite. This is a situation that the factory must prevent from happening. Thus our aim is to use a mathematical model to determine for which thicknesses of dust layers ignition occurs.

Irrigation

An important part of primary industry is the production of food on farms. In arid regions (e.g. in many parts of Australia), irrigation is often used to provide water for crops. In the case study of Chapter 6 we investigate the optimal size for irrigation furrows. The mathematical content involves the solution of a partial differential equation for the unsaturated flow of water in soils by assuming an expansion in trigonometric functions to take advantage of the periodicity of the problem.

Mathematical modelling to help understand complex processes

These case study problems involve many processes happening at once. Mathematical modelling will be used to consider only the **most important** physical processes. This, in turn, will allow us to obtain sufficiently simple equations on which we can make good mathematical progress.

This leads to a much better understanding of the more complicated system.

The ability to recast real-world problems in mathematical form is a remarkable fact of history. For a clear account of the steps involved in the process of mathematical modelling, we refer to Fulford et al. (1997), Edwards and Hamson (1989) and Fowkes and Mahony (1994). For a philosophical consideration of the apparently unreasonable effectiveness of mathematics in the physical sciences, the interested reader is referred to the classic article Wigner (1960).

1.4 Units and dimensions

In the physical world measured quantities are determined relative to some standard measurements. It is important that equations developed as part of our modelling process are consistent no matter which units are the basis of our measurements. This is called dimensional consistency.

Units

Units of a physical quantity are the reference measurements to which we make comparisons. Some examples are metres, minutes, joules, miles, kilograms, etc. The same quantity can be measured in different units (e.g. 1 km = 1,000 m = 0.6214 miles). In this example, each unit (kilometre, metre or mile) refers to a quantity described by length.

We call length a ***primary quantity***. Some other primary quantities are mass, time and temperature. ***Secondary quantities*** are those which are combinations of more than one primary quantity. For example, in the SI system velocity is measured in metres per second, which is a secondary quantity.

A variable which measures length is said to have ***dimension*** length, denoted by the symbol L. Thus a dimension L may take values of kilometre, metre or mile, depending on which system of units is adopted. Other dimensions, corresponding to some primary quantities, are mass, time and temperature, denoted by M, T and Θ respectively. The four primary units relevant to this book are listed below in Table 1.4.1. For a primary or secondary quantity q, $[q]$ denotes the dimensions of the quantity represented by the symbol q. The value of $[q]$ is expressed in terms of M, L, T and Θ.

Table 1.4.1. *Fundamental units of primary quantities.*

Primary Quantity	Symbol	SI Unit	cgs Unit
mass	M	kilogram, kg	gram, g
length	L	metre, m	centimetre, cm
time	T	second, s	second, s
temperature	Θ	kelvin, K	degree, °C

Other fundamental SI units include the ampere (A), the unit for electric current; the mole (mol), the unit for amount of a substance (i.e. the number of atoms or molecules); and the candela (cd), the unit for luminosity. All other units are derived from these base units and the ones in Table 1.4.1.

Rules for dimensions

Certain rules must be obeyed by a consistent set of units of measurement. They are mostly common sense. The rules are as follows:

(a) Two quantities may be **added** only if they have the **same dimensions**. Quantities of different dimensions may be multiplied or divided.

(b) Index Laws. If $[f] = M^{\alpha_1}L^{\alpha_2}T^{\alpha_3}\Theta^{\alpha_4}$ and $[g] = M^{\beta_1}L^{\beta_2}T^{\beta_3}\Theta^{\beta_4}$ then $[fg] = M^{\alpha_1+\beta_1}L^{\alpha_2+\beta_2}T^{\alpha_3+\beta_3}\Theta^{\alpha_4+\beta_4}$.

(c) Pure numbers are dimensionless, i.e. $[1] = 1$, $[2] = 1$, $[\pi] = 1$, $[0] = 1$. Thus multiplying by a pure number does not change the dimensions of a physical quantity, i.e. $[2m] = 1 \times M = M$.

(d) The dimensions of a derivative $\dfrac{\partial p}{\partial q}$ are $[p][q]^{-1}$. This is because a derivative is a limiting ratio of two quantities. Thus if u is temperature and x measures distance then $\left[\frac{\partial u}{\partial x}\right] = \Theta L^{-1}$. Also $\left[\frac{\partial^2 u}{\partial x^2}\right] = \Theta L^{-2}$, and more generally,

$$\left[\frac{\partial^{m+n}u}{\partial x^m \partial t^n}\right] = \Theta L^{-m}T^{-n}.$$

(e) The dimensions of an integral $\int_a^b p\,dq$ are given by $[p][q]$.

(f) The arguments of functions having Taylor expansions (of more than one term) must be dimensionless. This is because this is the only way we can add different powers of a quantity. For example, for

$$e^{kt} = 1 + kt + \frac{1}{2!}k^2 t^2 + \dots$$

where t is time, then $[k] = \mathtt{T}^{-1}$ since kt must be dimensionless.

A useful way of checking equations is to check they are ***dimensionally homogeneous*** . This means that both sides of an equation must have the same dimensions. The following example illustrates this.

Example 1: *Newton's second law gives*

$$F = ma \tag{1}$$

where F is the force on a particle, m is its mass and a is the acceleration of the particle. Check that equation (1) is dimensionally homogeneous.

Solution: *Force is measured in newtons which are $\mathrm{kg\,m\,s}^{-2}$. Thus $[LHS] = \mathtt{MLT}^{-2}$. Now $[m] = \mathtt{M}$ and $[a] = \mathtt{LT}^{-2}$. Thus $[RHS] = \mathtt{MLT}^{-2} = [LHS]$. So (1) is dimensionally homogeneous.*

Checking equations

Dimensions of secondary quantities can easily be obtained from the above rules. The following example shows how to do this.

Example 2: *Fourier's law is an equation relating heat flux to temperature gradient (see Section 1.6),*

$$J = -k\frac{\partial u}{\partial x},$$

where J is the heat flux, u the temperature, x denotes distance and k is the conductivity. Hence determine $[k]$.

Solution: *The heat flux, J, is heat energy per unit area per unit time. So*

$$[J] = \frac{[\text{energy}]}{[\text{area}][\text{time}]}.$$

Now energy has the dimensions of work, which is force times distance, so $[energy] = \mathtt{MLT}^{-2} \times \mathtt{L}$, *and* $[area] = \mathtt{L}^2$. *Hence*

$$[J] = \frac{\mathtt{M L^2 T^{-2}}}{\mathtt{L^2 T}}$$
$$= \mathtt{MT^{-3}}.$$

Now $[u] = \Theta$, *and* $[x] = \mathtt{L}$, *so*

$$\left[\frac{\partial u}{\partial x}\right] = \Theta \mathtt{L}^{-1}.$$

Since $[k] = [J] \times [\partial u / \partial x]^{-1}$ *then*

$$[k] = \mathtt{MLT^{-3}\Theta^{-1}}.$$

In SI units k is measured in $\mathrm{kg\,m\,s^{-3}\,K^{-1}}$. This is consistent with the above. For checking equations, Table 1.4.2 will be a useful reference.

Table 1.4.2. *Table of secondary quantities in mechanics and heat transport.*

Quantity	Dimensions	SI Units
density ρ	$\mathtt{ML^{-3}}$	$\mathrm{kg\,m^{-3}}$
velocity v	$\mathtt{LT^{-1}}$	$\mathrm{m\,s^{-1}}$
acceleration a	$\mathtt{LT^{-2}}$	$\mathrm{m\,s^{-2}}$
force F	$\mathtt{MLT^{-2}}$	newtons, N
pressure p	$\mathtt{ML^{-1}T^{-2}}$	$\mathrm{N\,m^{-2}}$, pascal, Pa
energy E	$\mathtt{ML^2T^{-2}}$	joule J
power $\dot{E}$	$\mathtt{ML^2T^{-3}}$	watt W
heat flux J	$\mathtt{MT^{-3}}$	$\mathrm{W\,m^{-2}}$
heat conductivity k	$\mathtt{MLT^{-3}\Theta^{-1}}$	$\mathrm{W\,m^{-1}\,K^{-1}}$
specific heat c	$\mathtt{L^2T^{-2}\Theta^{-1}}$	$\mathrm{J\,kg^{-1}\,K^{-1}}$
heat diffusivity α	$\mathtt{L^2T^{-1}}$	$\mathrm{m^2\,s^{-1}}$
Newton cooling coefficient h	$\mathtt{MT^{-3}\Theta^{-1}}$	$\mathrm{W\,m^{-2}\,K^{-1}}$

1.5 Diffusion equations

The derivation of the one-dimensional diffusion equation is based on the idea of mass conservation. In this section we give a detailed formulation of the 1-D diffusion equation.

Diffusion in a tube

Consider a circular tube. Let A be the cross-sectional area of the hollow part of the tube. The hollow part is filled with a mixture containing a *solute*. We assume the bulk mixture is *not moving* (but we consider this later in this section). However, if the solute concentration is higher at one end than at the other then the solute will diffuse towards the other end, as shown in Figure 1.5.1. We also assume the walls of the tube are impermeable to the solute.

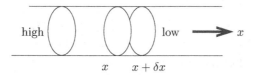

Fig. 1.5.1. Diffusion of a solute in a tube from high solute concentration to low solute concentration.

Let us define the concentration of the solute $C(x, t)$ as the ratio of the mass of the solute to the volume of the mixture. We can think of the concentration defined at a single point x by taking a small volume and then letting that volume tend to zero. Since the walls of the pipe are impermeable to the solute, the concentration of the solute will depend only on longitudinal position x and the time t.

We shall consider a small section x to $x + \delta x$ of the tube. As the solute diffuses through the tube the net change in the mass of the solute in the section is determined by the net difference in the mass of solute diffusing into and out of the tube. We can write this statement of conservation of mass as

$$\left\{\begin{array}{c} \text{rate of} \\ \text{change of} \\ \text{solute mass} \end{array}\right\} = \left\{\begin{array}{c} \text{net rate of} \\ \text{mass diffusing} \\ \text{in and out of section} \end{array}\right\}. \tag{1}$$

The term on the RHS refers to the net difference in rates of solute mass diffusing into the section and solute mass diffusing out of the section.

In terms of the concentration $C(x, t)$, the LHS of (1) can be written as the volume multiplied by the rate of change of concentration, evaluated at some internal point x_1, inside the section x to $x + \delta x$. Thus we write

$$\left\{\begin{array}{c} \text{rate of} \\ \text{change of} \\ \text{solute mass} \end{array}\right\} = A\delta x \frac{\partial C}{\partial t}(x_1, t) \tag{2}$$

where $A\delta x$ is the volume of the section.

Let us define $J(x, t)$ to be the mass flux of the solute, defined as the net rate of mass of solute diffusing through a cross-section at x, per unit cross-section area, per unit time. We can now write the diffusion in terms of the mass flux J.

$$\left\{ \begin{array}{c} \text{net rate of} \\ \text{mass diffusing} \\ \text{in and out of section} \end{array} \right\} = J(x, t)A - J(x + \delta x, t)A.$$

Hence, the mass balance equation (1) now becomes

$$A\delta x \frac{\partial C}{\partial t}(x_1, t) = A\delta x [J(x, t) - J(x + \delta x, t)].$$

Dividing by $A\delta x$, we obtain

$$\frac{\partial C}{\partial t}(x_1, t) = -\left[\frac{J(x + \delta x, t) - J(x, t)}{\delta x} \right].$$

We now let $\delta x \to 0$ and we thus obtain

$$\frac{\partial C}{\partial t} = -\frac{\partial J}{\partial x}, \tag{3}$$

using the definition of the partial derivative. Note that, as $\delta x \to 0$ we also have $x_1 \to x$, where $x < x_1 < x + \delta x$. Now **all** quantities are evaluated at the point x.

To relate the flux to the concentration we need a ***constitutive equation*** (an equation relating material variables, determined from experiments). The simplest one is ***Fick's law*** which states the mass flux is proportional to the concentration gradient. For 1-D diffusion, Fick's law can be written

$$J(x, t) = -D\frac{\partial C}{\partial x}(x, t) \tag{4}$$

where D is a positive constant known as the ***diffusivity***. Note the minus sign is included so that the solute diffuses in the direction of decreasing concentrations. Fick's law for diffusion is analogous to Fourier's law for heat conduction.

Substituting Fick's law (4) into the mass conservation equation (3) we obtain

$$\frac{\partial C}{\partial t} = D\frac{\partial^2 C}{\partial x^2} \tag{5}$$

which is known as the 1-D diffusion equation.

Advection

Advection is where the solute is carried along with the bulk movement of the fluid. We can think of the mass flux J to be due to both diffusion and advection, so $J = J_d + J_a$. The mass flux at position x is the rate of movement of mass per unit time per unit area through the cross-section at x.

Let $v(x,t)$ denote the fluid velocity (of the mixture). In the absence of diffusion the solute particles move at the same speed as the mixture. The total mass of solute that is transported through the cross-section is the volume of mixture moving past the cross-section in a time δt multiplied by the concentration. This volume is $vA\delta t$. Thus the mass flux due solely to advection is given by

$$J_a(x,t) = v(x,t)C(x,t).$$

Using Fick's law for the mass flux due only to diffusion of solute particles relative to the mean flow of the mixture, the total mass flux is given by

$$J(x,t) = v(x,t)C(x,t) - D\frac{\partial C}{\partial x}.$$

Substituting this into the mass conservation equation (3) we obtain the partial differential equation for the concentration

$$\frac{\partial C}{\partial t} = -\frac{\partial}{\partial x}\left(v(x,t)C(x,t) - D\frac{\partial C}{\partial x}\right)$$

which may be written as

$$\frac{\partial C}{\partial t} + \frac{\partial}{\partial x}(vC) = D\frac{\partial^2 C}{\partial x^2}. \tag{6}$$

If the moving mixture is an incompressible fluid then $v(x,t)$ is a constant. This follows from conservation of mass applied to the mixture — the mass flowing in ρAv, where ρ is the density of the mixture, must be constant. The previous equation (6) then simplifies to

$$\frac{\partial C}{\partial t} + v\frac{\partial C}{\partial x} = D\frac{\partial^2 C}{\partial x^2}. \tag{7}$$

Turbulent diffusion

So far we have thought about diffusion as due to random motion of molecules of a solute. However, the diffusion equation can occur in a wider context. In many air flows, especially on environmental scales, the velocity is turbulent. This means the velocity has a random component. Thus air pollution, for example, can be advected with the mean flow while simultaneously mixing with the air due to the random component of the air flow. This type of diffusion is called **turbulent diffusion**. In general this is a very complicated process that is not fully understood. In the simplest turbulent transport models, an eddy diffusivity is incorporated to relate turbulent flux to the gradient of mean concentration. (See Launder and Spalding (1972) for a more detailed discussion of the theory of eddy diffusivity. Wilcox (1994) and Weil (1988) give some extensions of this theory.) The value of the eddy diffusivity is usually several orders of magnitude larger than the diffusivity for molecular diffusion. In many problems it is typical for the diffusivity not to be constant. For example, the air becomes more turbulent with height from the ground.

For non-constant diffusivity, say $D(x)$ the governing equation for the concentration is not

$$\frac{\partial C}{\partial t} = D(x)\frac{\partial^2 C}{\partial x^2}.$$

A careful consideration of the derivation of the diffusion equation shows the appropriate form is

$$\frac{\partial C}{\partial t} = \frac{\partial}{\partial x}\left(D(x)\frac{\partial C}{\partial x}\right).$$

The generalised 1-D diffusion equation

We can consider the effects of advection, nonlinear diffusivity, and internal mass production. **Nonlinear diffusivity** occurs when the diffusivity depends on the concentration. **Internal mass production** is where the solute is created everywhere within the region of consideration (e.g. by some chemical reaction). Formulations of the modified diffusion equations for each of these phenomena are explored in the problems at the end of this chapter (see Question 6).

The results are summarised in the generalised 1-D diffusion equation

$$\frac{\partial C}{\partial t} + \frac{\partial}{\partial x}(vC) = \frac{\partial}{\partial x}\left(D(C)\frac{\partial C}{\partial x}\right) + M. \tag{8}$$

Here v is the bulk velocity of fluid flowing through the tube and the term M is the rate of production of solute, per unit time per unit volume. When the fluid motion is incompressible (so that the velocity v is independent of x) then the generalised 1-D diffusion equation simplifies to

$$\frac{\partial C}{\partial t} + v\frac{\partial C}{\partial x} = \frac{\partial}{\partial x}\left(D(C)\frac{\partial C}{\partial x}\right) + M.$$

The 3-D diffusion equation

A similar type of argument for mass transport yields the 3-D diffusion equation for concentration $C(\mathbf{x}, t)$,

$$\frac{\partial C}{\partial t} = D\nabla^2 C \qquad \text{where} \quad \nabla^2 = \frac{\partial^2}{\partial x^2} + \frac{\partial^2}{\partial y^2} + \frac{\partial^2}{\partial z^2} \tag{9}$$

in cartesian coordinates (x, y, z).

For 3-D problems, the generalised diffusion equation is

$$\frac{\partial C}{\partial t} + \boldsymbol{\nabla} \cdot (\mathbf{v}C) = \boldsymbol{\nabla} \cdot (D(C)\boldsymbol{\nabla}C) + M \tag{10}$$

where $D(C)$ is the concentration dependent diffusivity and M is the rate of production of mass of solute, per unit time per unit volume. For fluid flow which is incompressible ($\boldsymbol{\nabla} \cdot \mathbf{v} = 0$) the 3-D generalised diffusion equation simplifies to

$$\frac{\partial C}{\partial t} + \boldsymbol{\nabla} \cdot (\mathbf{v}C) = \mathbf{v} \cdot \boldsymbol{\nabla}C + M. \tag{11}$$

The reader who is familiar with fluid dynamics will recognise the advection term as coming from the **material derivative** (differentiation following the motion), see e.g. Acheson (1990).

1.6 Heat conduction equations

The fundamental equation describing heat conduction is a partial differential equation known as the **heat conduction equation** (or heat equation, for short). In this section we will see how to derive this equation for heat conduction along a long thin rod. The basic idea is that energy is conserved. We consider an infinitesimal section through the rod and account for the amount of heat energy entering and leaving the section. This rather simple approach can then be used on more complicated problems involving advection (heat carried along with a moving fluid), heat generation (e.g. by electrical resistance or chemical reaction), heterogeneity (different positions have different thermal properties) and nonlinearity (conductivity is temperature dependent).

Heat balance

Consider the heat flow in a **solid** rod with circular cross-section A. Assume that the surface of the rod is perfectly insulated so that no heat escapes radially. Thus the direction of heat flow is only in the longitudinal direction (along the axis of symmetry of the rod). Suppose that the rod is initially at a uniform low temperature. Then one end is suddenly raised to a higher temperature. Heat flows in the x-direction, from hot to cold, as shown in Figure 1.6.1.

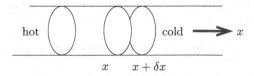

Fig. 1.6.1. Heat conduction in a rod.

Let δx be the thickness of a section through the rod located at the point x, where δx is taken to be very small compared to x. As the heat flows along the rod some of the heat will be absorbed by the rod as it raises the temperature of the rod. As a result of this, as heat flows into the cross-section at x a different amount of heat flows out at $x + \delta x$. We can write

$$\left\{ \begin{array}{c} \text{rate of} \\ \text{change of} \\ \text{heat content} \end{array} \right\} = \left\{ \begin{array}{c} \text{net rate of} \\ \text{heat conducted} \\ \text{in and out of section} \end{array} \right\} \tag{1}$$

by conservation of energy. This assumes that there is no heat production inside the rod or heat loss from the surface of the rod.

Formulating the equation

Let us now introduce some notation. Let $u(x,t)$ denote the temperature of the rod at position x at time t. Because there is no radial flow then the temperature will be constant over the cross-section provided it was constant initially.

Let us also define the **heat flux** $J(x,t)$ as the rate of heat passing through a cross-section, per unit area, per unit time. In terms of the heat flux, the term on the RHS of equation (1) becomes

$$\left\{ \begin{array}{c} \text{net rate of} \\ \text{heat conducted} \\ \text{in and out of section} \end{array} \right\} = J(x,t)A - J(x+\delta x,t)A. \qquad (2)$$

We now relate the LHS of equation (1) to the temperature. Some of the heat energy is absorbed by the rod and causes a change in the temperature of the rod. In a small time δt the temperature at x is changed by an amount $u(x,t+\delta t)-u(x,t)$ The amount of heat required to change the temperature of the entire mass of the section by this amount is proportional to both the mass of the section and the temperature difference. Thus

$$\left\{ \begin{array}{c} \text{rate of} \\ \text{change of} \\ \text{heat content} \end{array} \right\} = c\rho A \delta x\, \frac{\partial u}{\partial t}(x_1,t) \qquad (3)$$

where x_1 is some internal point $x < x_1 < x+\delta x$. Here $A\delta x$ is the volume of the section, ρ is the density and c is a proportionality factor called the specific heat. The specific heat is often taken to be constant, for a particular material, provided the temperature variation is not too great.

Substituting equations (3) and (2) into equation (1), and dividing by the product δx, we obtain

$$\rho c A \frac{\partial u}{\partial t} = -\frac{[J(x,t)-J(x+\delta x,t)]}{\delta x} A. \qquad (4)$$

Letting δx tend to zero we obtain, in the limit, (alternatively, take Taylor series of each of the terms)

$$\rho c \frac{\partial u}{\partial t} = -\frac{\partial J}{\partial x}. \qquad (5)$$

Note the minus sign on the RHS of equation (5). Equation (5) is the basic transport equation in one dimension. It needs to be supplemented by a constitutive equation which relates the heat flux J to the temperature u. For heat conduction we use Fourier's law.

Fourier's law

We now relate the RHS of equation (5) to the temperature. To a good approximation, for many solids, the heat flux is proportional to the temperature gradient. This is known as Fourier's law after the French mathematician and scientist, Fourier, who in 1822 published the first book on the mathematical theory of heat (and who is also famous for Fourier series and Fourier transforms).

Fourier's law may be written

$$J(x,t) = -k\frac{\partial u}{\partial x}(x,t) \tag{6}$$

where the proportionality factor k is known as the **thermal conductivity**. In some heat flow problems k can be a function of u, x or t. However, it is usual in mathematical modelling to make the simplest assumption initially; $k = $ constant. Later, we might relax this assumption if necessary.

Substituting Fourier's law (6) into the energy conservation equation (5) we obtain

$$\rho c \frac{\partial u}{\partial t} = k\frac{\partial^2 u}{\partial x^2},$$

taking the conductivity k to be constant.

This equation is often written in the form

$$\frac{\partial u}{\partial t} = \alpha\frac{\partial^2 u}{\partial x^2}, \qquad \alpha = \frac{k}{\rho c}, \tag{7}$$

known as the **heat equation**. Note the similarity in form to the diffusion equation from Section 1.5. Here the constant $\alpha = k/\rho c$ is called the **heat diffusivity**. It characterises the ability of heat energy to diffuse through a given material. The heat equation (7) is a partial differential equation in two independent variables, time t, and position x.

Extensions of the heat equation

Modifications of the basic heat equation include heat sources, advection, and temperature dependent conductivity. A generalised heat equation, incorporating all of these is

$$\rho c \left(\frac{\partial u}{\partial t} + \frac{\partial}{\partial x}(vu) \right) = \frac{\partial}{\partial x}\left(k(u)\frac{\partial u}{\partial x} \right) + Q \qquad (8)$$

where $v(x,t)$ is the speed of a fluid, Q is the rate of heat produced internally in the fluid, per unit volume, per unit time and $k(u)$ is the temperature dependent conductivity. If the fluid flow is incompressible, then v is independent of x, and the generalised 1-D heat equation simplifies to

$$\rho c \left(\frac{\partial u}{\partial t} + v\frac{\partial u}{\partial x} \right) = \frac{\partial}{\partial x}\left(k(u)\frac{\partial u}{\partial x} \right) + Q.$$

A heat source can arise through an exothermic reaction within the material which generates heat at every point. It can also arise from an electric current which creates heat within the material as it experiences electrical resistance. Another way heat can be generated within a material is by nuclear fission in a nuclear fuel rod.

Advection is where heat is transported due to bulk movement of a fluid. This is sometimes referred to as convection in the context of heat. (See Section 1.5 for an explanation of advection for mass transport.) Suppose we have a fluid which is moving with velocity $v(x,t)$. In this case heat is transported both by conduction and by advection, due to bulk movement of material and also due to random movement of molecules. The term $\rho c \partial(vu)/\partial x$ in the above equation arises due to advection. When the movement of fluid is due to buoyancy (with hot fluid being less dense than cold fluid and therefore rising), the heat transport is known as convection.

The conductivity of a material will often vary with temperature, if the range of temperatures is large. The functional dependence of k on u will be determined experimentally. Note that, in equation (8), the term $k(u)$ must stay inside the $\partial/\partial x$ term, unless k is constant with respect to u.

The 3-D heat conduction equation

If the temperature depends on more than one spatial variable then we need to account for heat fluxes in different directions. This is accomplished by defining a vector valued heat flux.

We can define the vector heat flux $\mathbf{J}(\mathbf{x}, t)$. The direction of the vector is the direction of heat flow at the point $\mathbf{x}$, and its magnitude is the rate of heat flow per unit area, per unit time.

We must generalise Fourier's law to three dimensions. Recall that the 1-D version of Fourier's law stated that the heat flux was proportional to the temperature gradient. Thus in three dimensions

$$\mathbf{J}(\mathbf{x}, t) = -k\boldsymbol{\nabla} u \tag{9}$$

where k is the thermal conductivity. This gives the direction of heat flow as the direction of maximum rate of decrease of temperature (since from vector calculus the gradient gives the direction of maximum rate of increase). Also, we have assumed that the material is *isotropic* which means that there is no preferred direction of heat flow within the material itself. If a material were non-isotropic then (9) would need to be generalised to a linear combination of the partial derivatives of the temperature.

By assuming conservation of heat energy for an arbitrary region, and using the divergence theorem, it is possible to derive the 3-D heat equation (details are left to the exercises, see Question 8 for the mass diffusion case)

$$\rho c \frac{\partial u}{\partial t} + \boldsymbol{\nabla} \cdot \mathbf{J} = 0. \tag{10}$$

Substituting Fourier's law (9) into the transport equation (10) we obtain

$$\rho c \frac{\partial u}{\partial t} = k\nabla^2 u \qquad \text{where} \quad \nabla^2 = \frac{\partial^2}{\partial x^2} + \frac{\partial^2}{\partial y^2} + \frac{\partial^2}{\partial z^2}.$$

Dividing by ρc gives

$$\frac{\partial u}{\partial t} = \alpha \nabla^2 u, \qquad \text{where} \quad \alpha = \frac{k}{\rho c}. \tag{11}$$

This is the three-dimensional generalisation of the linear heat equation.

Note that for one-dimensional problems, where u is a function of only x and t then equation (11) reduces to the one dimensional heat equation.

Just as we generalised the 1-D heat equation to account for nonlinear conduction, heat sources, advection and heterogeneity, we can do the same for the 3-D heat equation. The generalised version is

$$\rho c \left(\frac{\partial u}{\partial t} + \boldsymbol{\nabla} \cdot (\mathbf{v} u) \right) = \boldsymbol{\nabla} \cdot (k(u) \boldsymbol{\nabla} u) + Q, \qquad (12)$$

where $\mathbf{v}$ is the bulk velocity of the material, $k(u)$ is the temperature dependent conductivity and Q is the rate of production of heat per unit time per unit volume. For an incompressible fluid ($\boldsymbol{\nabla} \cdot \mathbf{v} = 0$) the equation simplifies to

$$\rho c \left(\frac{\partial u}{\partial t} + \mathbf{v} \cdot \boldsymbol{\nabla} u \right) = \boldsymbol{\nabla} \cdot (k(u) \boldsymbol{\nabla} u) + Q. \qquad (13)$$

1.7 Boundary conditions

We investigate some of the common boundary conditions for heat conduction problems. There are analogous boundary conditions for mass-diffusion problems.

Number of boundary and initial conditions

The classical heat-diffusion equation for heat flow in a rod is

$$\frac{\partial u}{\partial t} = \alpha \frac{\partial^2 u}{\partial x^2} \qquad (1)$$

which is a second-order partial differential equation in space and a first-order partial differential equation in time.

We require one initial condition (e.g. an initial distribution of temperature) and two boundary conditions — one at each end of the rod. For the three-dimensional heat equation

$$\frac{\partial u}{\partial t} = \alpha \nabla^2 u$$

we require an initial condition plus data specified on the boundary of the domain of the problem. There are several common forms of boundary conditions which occur in practical problems. We list some of these below.

A. Prescribed temperature boundary condition

This is probably the simplest type of boundary condition. Here the temperature on some boundary is a specified function of time — usually a constant, although it may be a given function of time. An example is

$$u(L,t) = u_1 \tag{2}$$

where u_1 is a constant. This might occur when one end of a heated pipe is immersed in a large bath of ice water at temperature say $0\,°C$, so $u_1 = 0$ in (1). In the mathematics literature, this is referred to as a **Dirichlet boundary condition**.

B. Prescribed heat flux boundary condition

Another boundary condition is that in which the rate of heat flow at a boundary is known. This is usually given in terms of a prescribed heat flux, e.g.

$$J(L,t) = -k\frac{\partial u}{\partial x}(L,t) = J_1 \tag{3}$$

where k is the conductivity. Note that J_1 describes the prescribed heat flux density through a cross-section in the x-direction. As an example, consider a solar hot-water heater. Suppose there is an inflow of 10 watt/square metre, (in the opposite direction to the x-axis). Then the boundary condition at $x = L$ is given by

$$J(L,t) = -k\frac{\partial u}{\partial x}(L,t) = -10\,\text{watt}$$

where the minus sign indicates that the flow of heat is in the negative x direction. In the mathematics literature, this is referred to as a **Neumann boundary condition**.

A special case of the prescribed flux boundary condition occurs for perfect insulation where no heat can get through the boundary. Then equation (3) takes the form

$$J(L,t) = 0.$$

Using Fourier's law, $J = -k\partial u/\partial x$, this can also be written as

$$\frac{\partial u}{\partial x}(L,t) = 0. \tag{4}$$

C. Newton cooling boundary condition

Another boundary condition, which accounts for imperfect insulation, is **Newton's law of cooling**, also called convective cooling. This is an empirical model rather than a law and states that the rate of heat loss flux density is proportional to the difference in temperature of the material and its surroundings.

This is expressed as

$$J(L,t) = \pm h[u(L,t) - u_s]$$

where the proportionality factor h is called the heat transfer coefficient and u_s is the temperature of the surroundings. The appropriate sign is chosen to give the correct sign of J in any given problem. Using Fourier's law

$$-k\frac{\partial u}{\partial x}(L,t) = \pm h[u(L,t) - u_s].$$

In the mathematics literature, this is sometimes referred to as a **Robin boundary condition**.

The value of h depends on the material, the type of material of the surroundings and on the velocity of the fluid flowing past in the surroundings. Thus strong cold winds lead to more rapid heat loss and consequently higher values of h. This is called a "wind chill" factor.

In the case $h = 0$ we recover the perfect insulation condition (4). Also as $(1/h) \to 0$ we recover the prescribed temperature condition (2).

D. Radiation boundary condition

Heat transfer by radiation involves transfer of energy by electromagnetic waves, the interaction of these waves with a body causing it to be heated. Here, the boundary condition is given by

$$J(L,t) = -k\frac{\partial u}{\partial x} = \epsilon\sigma[u^4(L,t) - u_s^4] \tag{5}$$

where $\sigma = 5.6697 \times 10^{-8}\,\mathrm{W\,m^{-2}\,K^4}$ is the Stefan–Boltzmann constant, ϵ, $(0 < \epsilon < 1)$, is a proportionality factor called the **emissivity** and u_s is the background temperature of the surroundings. Here the temperatures $u(x,t)$ and u_s **must** be measured in kelvin, not degrees Celsius. For further information on radiative transfer, the interested reader is referred to Holman (1992).

Note that equation (5) is a nonlinear boundary condition. Because of this it is only used in numerical work. In environmental applications u may be considered close to u_s (e.g. in a typical winter u may vary by only 6% from 273 K to 290 K) and the boundary condition (5) may be approximated by the first linear term of the Taylor series

$$f(u) = f'(u_s)(u - u_s) = 4\epsilon\sigma u_s^3(u - u_s).$$

This is mathematically equivalent to Newton's Law of cooling with convective heat transfer coefficient $h = 4\epsilon\sigma u_s^3$.

E. Continuity boundary condition

Another type of boundary condition that occurs in heat conduction problems is the thermal contact between two different media. We say that the media are in **perfect contact** if they are welded tightly so that heat flows smoothly between them. We then assume that the temperature $u(x,t)$ and the heat flux density $J(x,t)$ are continuous across the boundary, i.e.

$$u_1(L,t) = u_2(L,t),$$
$$J_1(L,t) = J_2(L,t),$$

where u_1 and J_1 correspond to the temperature and heat flux on one side of the boundary $x = L$ and u_2 and J_2 are the temperature and heat flux on the other.

F. Moving boundaries

Moving boundaries occur in problems involving melting and solidification. These are more complicated since we do not know the actual location of the boundary before solving the problem. We will see how to formulate these types of boundary condition in Chapter 2.

An example

As an example of determining appropriate boundary conditions consider a blast furnace wall consisting of a layer of brick (conductivity k_1) and an outer layer of asbestos (conductivity $k_2 < k_1$). This is shown in Figure 1.7.1. Suppose that the outside of the outer wall radiates heat, modelled by Newton's law of cooling into the surrounding air, at temperature U_3 and the inside of the inner wall is maintained at temperature U_0.

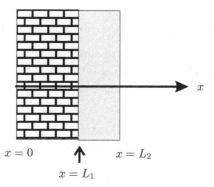

$$x = 0 \qquad \uparrow \qquad x = L_2$$
$$x = L_1$$

Fig. 1.7.1. A composite blast furnace wall, comprising a brick wall and an asbestos wall. The temperature on the inner wall $x = 0$ is U_0 and heat is lost to the air (at temperature U_3) from the other wall $x = L_2$ according to Newton's law of cooling.

The governing equations for one-dimensional flow are the heat equations for the two regions,

$$\rho_1 c_1 \frac{\partial u_1}{\partial t} = k_1 \frac{\partial^2 u_1}{\partial x^2}, \qquad 0 < x < L_1,$$

$$\rho_2 c_2 \frac{\partial u_2}{\partial t} = k_2 \frac{\partial^2 u_2}{\partial x^2}, \qquad L_1 < x < L_2.$$

Example 1: *Give a reasonable set of boundary conditions for the temperatures u_1 and u_2, corresponding to Figure 1.7.1.*

Solution: *The appropriate system of boundary conditions is*

$$u_1(0, t) = U_0,$$
$$u_1(L_1, t) = u_2(L_1, t),$$

which expresses continuity of temperature across $x = L_1$,

$$-k_1 \frac{\partial u_1}{\partial x}(L_1, t) = -k_2 \frac{\partial u_2}{\partial x}(L_1, t),$$

which expresses continuity of heat flux across $x = L_1$, and on $x = L_2$ we have the Newton cooling condition

$$-k_2 \frac{\partial u_2}{\partial x} = h[u_2(L_2, t) - U_3].$$

Boundary conditions for 3-D transport

Generalising boundary conditions to three dimensions is quite straight forward. The main thing to remember is that the heat flux *into* a surface is given by $-\mathbf{J} \cdot \hat{\mathbf{n}}$ where $\hat{\mathbf{n}}$ is the unit *outward* normal to the bounding surface.

For example, for the problem of heat flux J_1 into a solid sphere, then the appropriate boundary condition on the sphere surface $r = a$ is

$$-\mathbf{J} \cdot \hat{\mathbf{n}} = J_1$$

which reduces to

$$k \frac{\partial u}{\partial r} = J_1$$

after expressing the radial component of $\mathbf{J}$ in spherical coordinates with radial symmetry (see, for example, Spiegel (1968)), noting that $\nabla u = (\partial u / \partial r)\hat{\mathbf{e}}_r$ with $\hat{\mathbf{e}}_r$ as the unit outward normal vector).

Diffusion boundary conditions

The boundary conditions for diffusion problems take the same form as some of the boundary conditions for heat conduction. For example, if the concentration is prescribed on a boundary then the boundary condition is given by (2) (with u replaced by C). If the boundary is impermeable (no solute or solution can cross it) then the boundary condition is the zero-flux condition, describing zero mass transport, $J(L, t) = 0$.

1.8 Solving the heat/diffusion equation

We look at finding equilibrium solutions for the 1-D heat and diffusion equations. This leads us to solve an ordinary differential equation. We then review briefly the techniques that will be introduced in the following chapters for solving time-dependent partial differential equations.

Equilibrium solution

With an appropriate set of boundary conditions and an initial condition we can, in principle, solve the diffusion equation. In practice it is necessary to choose the appropriate technique.

If we are not interested in the time dependence of the solution then it is usually not difficult to obtain a steady-state solution, where we neglect the time derivative. For 1-D diffusion equations this results in an ordinary differential equation.

Example 1: *Find the equilibrium solution for the partial differential equation*

$$\frac{\partial C}{\partial t} = D\frac{\partial^2 C}{\partial x^2} - C$$

with the boundary conditions

$$\frac{\partial C}{\partial x} = 0, \qquad C(1,t) = 1.$$

Solution: *For the equilibrium solution we neglect the time dependence, writing the concentration $C(x,t)$ as $C(x)$ and setting $\partial C/\partial t = 0$. The equilibrium solution then satisfies*

$$\frac{d^2 C}{dx^2} - C = 0,$$

with boundary conditions

$$\frac{dC}{dx}(0) = 0, \qquad C(1) = 0.$$

The general solution is $C(x) = A\sinh(x) + B\cosh(x)$. Applying the boundary conditions gives $A = 0$ and $B = 1/\cosh(1) = \mathrm{sech}(1)$. Hence the equilibrium solution is

$$C(x) = \frac{\cosh(x)}{\cosh(1)}.$$

Overview of techniques

Equilibrium solutions to time-dependent partial differential equations are useful in that they tell us what the long term behaviour of the solution is (the steady-state). However, information about how the solution evolves into the steady-state is usually obtained by solving the full time-dependent partial differential equation. This can sometimes be difficult.

In the following chapters we introduce the reader to a number of techniques for solving (or approximately solving) the 1-D heat/diffusion equation. Each of these techniques is motivated by a relevant case study from industry. The techniques are:

- **The Boltzmann similarity solution.** (Chapter 2). Here we use the substitution $u = f(x/\sqrt{t})$ to reduce the standard 1-D heat/diffusion equation to an ordinary differential equation. It is applicable to the basic heat or diffusion equation on a semi-infinite domain. We also see how to use this method to solve a problem with a moving boundary. and show how to obtain a pseudo-steady-state approximate solution.

- **The method of stretching transformations.** (Chapter 3). This is an extension of the technique use in Chapter 2, where we make a substitution corresponding to a stretching of each of the variables and use this to come up with a similarity substitution which reduces the partial differential equation to an ordinary differential equation.

- **The method of perturbations.** (Chapter 4). Sometimes a problem that is difficult to solve in general may be close, in some sense, to a problem which is easier to solve. We develop the method of regular perturbations which gives a systematic way of obtaining a sequence of approximate equations. The technique involves substituting a power series expanded about some small parameter.

- **Bifurcation analysis.** (Chapter 5). Solutions to problems will involve parameters. For some nonlinear problems the solution can exhibit sudden changes (or jumps) as those parameters change. Bifurcation analysis, as introduced here, is essentially a graphical technique, where the equilibrium solution is plotted with respect to a parameter.

- **The method of Fourier expansions.** (Chapter 6). For some problems there is a natural periodic structure to the problem. Provided the governing equations are linear, we can sometimes take advantage of the periodic structure and look for solutions as a series of trigonometric functions. This is called a Fourier series. (It can also be used for problems on finite domains by pretending the problem is replicated over space.)

1.9 Scaling equations

When modelling a complicated physical system, in the early stages of model development, it is desirable to neglect the least important factors in the problem. This enables us to develop a less ambiguous causal relationship between the most important parameters of a system and the outcome of an experiment.

Uses of scaling

Using dimensional analysis (see Section 1.10) tells us which dimensionless parameters are important in a problem. It is also useful to see how these dimensionless parameters manifest themselves in the governing differential equations and boundary conditions. The process of making differential equations and boundary conditions dimensionless is called *scaling*.

Thus if the value of the dimensionless parameter is very small (or very large) we may be able to neglect one or more terms in the equations, perhaps obtaining significant simplification of the problem.

Example problem

As an example of the technique of scaling equations we consider the one dimensional heat equation for a rod of length ℓ with a prescribed temperature at one end $(x = 0)$ and Newton cooling at the other end $(x = \ell)$ into surroundings at zero temperature.

The governing equation is the heat equation

$$\frac{\partial u}{\partial t} = \alpha \frac{\partial^2 u}{\partial x^2} \tag{1}$$

where α is the heat diffusivity.

At one end of the rod we have the prescribed temperature boundary condition,

$$u(0, t) = u_1, \tag{2}$$

and, at the other end, the Newton cooling boundary condition,

$$-k\frac{\partial u}{\partial x} = hu(\ell, t) \tag{3}$$

where k is the conductivity and h is the convective heat transfer coefficient.

Dimensionless variables

The basic idea is to choose constants which have the same dimensions as the variables in the equations (we call these scales). We then obtain dimensionless variables by dividing the original variables by the scales. These are then substituted into the governing equations, the boundary conditions and the initial conditions.

The parameters of the problem are ℓ, α, u_1, k and h. We use these to find the scales for the variables u, x and t. Scales for x and u are easy to find. In both these cases constants exist which have the same dimensions as the variables x and u. We can use ℓ for x and u_1 for u. There is no single parameter with dimensions of time to use as a scale for t so we need to construct one from the available parameters. One possibility is the combination ℓ^2/α which has dimensions $[\ell^2/\alpha] = \mathrm{T}$.

We now introduce **dimensionless ratios** of the variables. We define the dimensionless length X, dimensionless time T, and dimensionless temperature U by

$$X = \frac{x}{\ell}, \qquad T = \frac{\alpha t}{\ell^2}, \qquad U = \frac{u}{u_1}. \tag{4}$$

Example 1: *Using the chain rule and the change of variables (4), express* $\dfrac{\partial u}{\partial x}$ *and* $\dfrac{\partial^2 u}{\partial x^2}$ *in terms of the dimensionless variables U and X.*

Solution: *From equation (4)*

$$x = \ell X, \qquad t = \frac{\ell^2}{\alpha}T, \qquad u = u_1 U. \tag{5}$$

By the chain rule,

$$\frac{\partial u}{\partial x} = \frac{\partial}{\partial X}(u_1 U) \times \frac{dX}{dx} = \left(\frac{u_1}{\ell}\right)\frac{\partial U}{\partial X}.$$

Hence, using the chain rule again,

$$\frac{\partial^2 u}{\partial x^2} = \frac{\partial}{\partial x}\left(\frac{\partial u}{\partial x}\right) = \frac{\partial}{\partial X}\left(\frac{u_1}{\ell}\frac{\partial U}{\partial X}\right) \times \frac{dX}{dx} = \left(\frac{u_1}{\ell^2}\right)\frac{\partial^2 U}{\partial X^2}.$$

Let us look at the pattern of what is happening. The effect on the **dependent variable** u is to multiply by its scale u_0. The effect on the independent variable x is to divide by its scale ℓ, for each order of differentiation. In general, if

$$U = u_0 u, \qquad X = x_0 x, \qquad T = t_0 t$$

then

$$\frac{\partial^{m+n} u}{\partial x^m \partial t^n} = \left(\frac{u_0}{x_0^m t_0^n}\right)\frac{\partial^{m+n} U}{\partial X^m \partial T^n}. \tag{6}$$

Example 2: Use the formula (6) to express the derivatives $\partial u/\partial t$ and $\partial^2 u/\partial x^2$ in terms of the dimensionless variables U, X and T.

Solution: By the formula, letting $t_0 = \ell^2/\alpha$, $x_0 = \ell$, $u_0 = u_1$ and taking $m = 0$, $n = 1$,

$$\frac{\partial u}{\partial t} = \frac{u_0}{\ell^2/\alpha}\frac{\partial U}{\partial T}.$$

With $m = 2$, $n = 0$ and $x_0 = \ell$,

$$\frac{\partial^2 u}{\partial x^2} = \frac{u_0}{\ell^2}\frac{\partial^2 U}{\partial X^2}.$$

With practice, you should be able to apply the formula (6) very quickly and not have to use the chain rule to scale derivatives.

Scaling the partial differential equation

We now use the change of variable (5) to replace the dimensional variables in (1) with non-dimensional ones. In principle, we use the chain rule, but in practice we use the formula (6). Since the relationship between variables is simply multiplying by a constant, the change of variables is simple.

Example 3: Using (4) scale the partial differential equation (PDE), equation (1). That is, express this PDE in dimensionless variables U, X and T.

Solution: From Example 1 or Example 2 the PDE (1) becomes

$$\left(\frac{u_1}{\ell^2/\alpha}\right)\frac{\partial U}{\partial T} = \alpha\left(\frac{u_1}{\ell^2}\right)\frac{\partial^2 U}{\partial X^2}.$$

This simplifies to

$$\frac{\partial U}{\partial T} = \frac{\partial^2 U}{\partial X^2}.$$

Scaling boundary conditions

We now turn our attention towards scaling the boundary conditions . It is useful to first write out exactly what the boundary condition means. For example, the boundary condition $u(0,t) = u_1$ can be written

$$u = u_1 \quad \text{when } x = 0.$$

Then we simply replace the lower case variables with the upper case variables, using (4).

Example 4: *Scale both the boundary conditions*

$$u(0,t) = u_1, \qquad -k\frac{\partial u}{\partial x}(\ell,t) = hu(\ell,t).$$

using the change of variable (5).

Solution: The two boundary condition can be written out as

$$u = u_1 \quad \text{when } x = 0, \qquad -k\frac{\partial u}{\partial x} = hu \quad \text{when } x = \ell.$$

Using (4) these become

$$u_1 U = u_1 \quad \text{when } \ell X = 0, \qquad -\frac{ku_1}{\ell}\frac{\partial U}{\partial X} = hu_1 U \qquad \text{when } \ell X = \ell.$$

These simplify to

$$U = 1 \quad \text{when } X = 0, \qquad -\frac{\partial U}{\partial X} = \epsilon U \quad \text{when } X = 1$$

where

$$\epsilon = \frac{h\ell}{k}.$$

Since U is defined to be a function of the dimensionless variables X and T, we can hence write the scaled boundary conditions as†

$$U(0,T) = 1, \qquad -\frac{\partial U}{\partial X}(1,T) = \epsilon U(1,T).$$

Summary: the scaled problem

In summary, we have taken the problem

$$\frac{\partial u}{\partial t} = \alpha \frac{\partial^2 u}{\partial x^2}, \qquad u(0,t) = u_1, \qquad -k\frac{\partial u}{\partial x}(\ell,t) = hu(\ell,t)$$

and written it in terms of dimensionless variables as

$$\frac{\partial U}{\partial T} = \frac{\partial^2 U}{\partial X^2}, \qquad U(0,T) = 1, \qquad -\frac{\partial U}{\partial X}(1,T) = \epsilon U(1,T).$$

The scaled problem is simpler to work with — it only involves **one** dimensionless parameter ϵ, whereas the original equations involved **five** parameters, α, h, k, ℓ and u_1. Note that each term in each equation is itself dimensionless.

A quick check of the dimensions shows that the parameter

$$\epsilon = \frac{h\ell}{k}$$

is dimensionless. Physically, it gives us a measure of the relative importance of heat lost from the boundary. In the literature, this parameter is called the **Biot number.**

1.10 Dimensional analysis

Dimensional analysis is a technique used to reduce a problem to a minimal set of dimensionless variables. It is a useful technique for simplifying a problem. The general theory of dimensional analysis may be found in the classic texts by Barenblatt (1987) and Birkhoff (1950) with a good introduction given by Lin and Segel (1974). Here we give the basic idea, apply the technique to some examples and then summarise the general principles.

† Note that it is not correct to write $U(0, \ell^2 t/\alpha) = 1$.

The basic idea

Before any serious modelling with differential equations begins it is often useful to get a feel for the important variables in a problem by using the technique of dimensional analysis. This technique involves finding the simplest possible functional relationships between the important variables of the problem. The following example illustrates this type of problem.

Any given problem will depend on a number of variables and parameters. The variables will often be quantities representing position and time. Parameters are physical constants in a problem. For example, in problems involving heat conduction, parameters may include thermal material constants such as conductivity and specific heat, and also constants referring to the size of an object, such as its overall length.

An example problem

Let us consider a simple problem involving heat conduction.

Consider a semi-infinite rod defined on $0 < x < \infty$ as a model for a very long rod which is heated at one end. The initial temperature of the rod is $0\,°C$ and one end of the rod ($x = 0$) is subsequently exposed to a constant temperature u_1. We assume that there is no heat loss from the surface of the rod. Hence we have a one-dimensional heat conduction problem.

The temperature u will depend on the distance x from the end of the rod and also on the time t. Since the rod is of infinite length there is no length scale for the problem. The temperature also will depend on thermal properties. A single constant, called the thermal diffusivity, α, characterises the material. The dimensions of α are L^2T^{-1}. We express the dependence of the temperature on these variables and parameters by writing

$$u = f(x, t, \alpha, u_1). \tag{1}$$

Applying dimensional analysis

Our aim is to express (1) as a relationship among **dimensionless quantities**. We start by taking the dimensions of (1):

$$[u] = [f(x, t, \alpha, u_1)]. \tag{2}$$

In general, for a function of several variables, the dimensions of the value of a function f will be the product of all the variables, each raised to some arbitrary power. Thus, for example, equation (2) can be expressed as

$$[u] = [x]^a [t]^b [\alpha]^c [u_1]^d \tag{3}$$

where a, b, c and d are arbitrary constants to be determined. In general, if we have a function $f(p_1, p_2, p_3, \ldots)$ then

$$[f(p_1, p_2, p_3, \ldots)] = [p_1]^a [p_2]^b [p_3]^c \ldots.$$

We can now determine these constants by requiring the dimensions of both sides of the equation to be the same. The following example shows how to do this.

Example 1: *Express equation (3) in terms of dimensions* L, T *and* Θ. *Hence reduce the problem to the minimal number of variables.*

Solution: *Now* $[x] = \mathsf{L}$, $[t] = \mathsf{T}$, *and* $[u] = [u_1] = \Theta$ *and* $[\alpha] = \mathsf{L}^2\mathsf{T}^{-1}$. *Hence, equating dimensions in (1) gives*

$$\Theta = \mathsf{L}^a \mathsf{T}^b (\mathsf{L}^2 \mathsf{T}^{-1})^c \Theta^d$$
$$= \mathsf{L}^{a+2c} T^{b-c} \Theta^d.$$

Equating the exponents of each fundamental dimension gives a set of simultaneous linear equations

$$a + 2c = 0, \quad b - c = 0, \quad d = 1.$$

Solving this system we obtain

$$b = -\frac{1}{2}a, \qquad c = -\frac{1}{2}a, \qquad d = 1.$$

The system is **underdetermined** *and we have expressed the solution in terms of* **one** *unknown a. Hence*

$$[u] = [u_1][x]^a [t]^{-a/2} [\alpha]^{-a/2}$$

which simplifies to

$$[u] = [u_1] \left[\frac{x}{\sqrt{\alpha t}} \right]^a . \tag{4}$$

We now go back to expressing u as a functional relationship — this time of the dimensionless variables. Since there is only one arbitrary constant a, then the function will be of only one variable. Equation (4) is consistent with the the functional relationship

$$u = u_1 F(\frac{x}{\sqrt{\alpha t}}) \qquad (5)$$

for some function F of **one** variable.

In fact what (5) says is that the dimensionless combination u/u_1 only depends on one dimensionless combination $x/\sqrt{\alpha t}$. This idea can be very useful in solving certain partial differential equations. It allows us to effectively reduce the number of independent variables by one, converting a partial differential equation in two variables to an ordinary differential equation in one variable (see Section 2.2 for how this works with the 1-D heat conduction equation).

Summary of the technique

In dimensional analysis the main idea is to start with a general functional relationship between various physical variables and then to obtain a simplified relationship based on dimensionless variables. The main skill lies in deciding which variables are important in a problem, the rest is routine.

Given a functional relationship of the form

$$\phi = f(x_1, x_2, \ldots, x_n)$$

first take the dimensions of both sides. This yields something of the form

$$[\phi] = [x_1]^{a_1} [x_2]^{a_2} \ldots [x_n]^{a_n}.$$

By reducing each of the quantities $[x_i]$, $i = 1 \ldots n$ to fundamental dimensions L, M, T, and Θ and equating with the LHS, a linear system of equations should be obtained for the unknowns $a_1, \ldots, a_n$.

Solving the linear system yields several possibilities.

- *Unique solution:* Solve for the unknowns. Then ϕ is given by a dimensionless constant multiplied by the variables raised to the powers corresponding to the unique solutions obtained for $a_1, \ldots, a_n$.

- *More (independent) equations than unknowns:* Solve for some of the variables in terms of the others, reduced as far as possible. The number of free parameters gives the number of independent dimensionless variables that the solution depends on.

- *Fewer (independent) equations than unknowns:* This shows that there are not enough variables in the original functional relationship. Think about the problem again and come up with some additional variables.

The following examples illustrate the last two situations.

Example 2: *Consider the same problem of the rod heated at one end at temperature u_1 and at the other end, $x = \ell$, no heat escapes. Use dimensional analysis to express the temperature in terms of a minimal number of dimensionless quantities.*

Solution: *The temperature is now a function of the variables x and t and the constants ℓ and α. So we can write*

$$u = f(x, t, \alpha, \ell).$$

Taking dimensions of the equation, we have

$$[u] = [x]^a [t]^b [\alpha]^c [\ell]^d [u_1]^e \tag{6}$$

where a, b, c, d and e are constants.

Evaluating the dimensions of each of the quantities u, x, t, α and ℓ, we obtain

$$\Theta = L^a T^b (L^2 T^{-1})^c L^d \Theta^e$$

which simplifies to

$$\Theta = L^{a+2c+d} T^{b-c} \Theta^e.$$

We thus obtain the equations

$$a + 2c + d = 0, \quad b - c = 0, \quad e = 1.$$

So $c = b$ and $d = -a - 2b$ and we still have a and b undetermined.

Putting this back into equation (6) we obtain

$$[u] = [x]^a [t]^b [\alpha]^b [\ell]^{-a-2b} [u_1],$$

which we can also write as

$$u = [u_1] \left[\frac{x}{\ell} \right]^a \left[\frac{\alpha t}{\ell^2} \right]^b.$$

Since there are two terms with arbitrary constants a and b this tells us that the functional form must involve a function of **two variables**. Thus we can write

$$u = u_1 F\left(\frac{x}{\ell}, \frac{\alpha t}{\ell^2}\right).$$

Suppose we had forgotten to include a variable. Dimensional analysis can tell us we have made an error — we have left something important out. The following example shows how this comes about.

Example 3:   *Consider an infinite rod with no heat loss from the curved surface. Use dimensional analysis on this problem where the temperature u_1 of the end of the rod has not been included.*

Solution: *We have*

$$u = f(x, t, \alpha).$$

Taking dimensions gives

$$[u] = [x]^a [t]^b [\alpha]^c.$$

Expression in terms of fundamental quantities L, T, *and* Θ,

$$\Theta = \mathrm{L}^a \mathrm{T}^b (\mathrm{L}^2 \mathrm{T}^{-1})^c.$$

Equating like terms gives

$$a + 2c = 0, \quad b - c = 0, \quad 1 = 0.$$

This is not consistent. The original assumption regarding the functional dependence of variables must have been wrong.

1.11 Problems for Chapter 1

1.   *For the following determine the dimensions of the required variable.*

(a) *Given* $[k] = \mathrm{MLT}^{-3}\Theta^{-1}$, $[\rho] = \mathrm{ML}^{-3}$ *and* $[c] = \mathrm{L}^2\mathrm{T}^{-2}\Theta^{-1}$ *find* $[\alpha]$ *where* $\alpha = k/(\rho c)$.

(b) *Verify that the quantity* hL/k *is dimensionless. You are given* $[h] = \mathrm{MT}^{-3}\Theta^{-1}$, $[k] = \mathrm{MLT}^{-3}\Theta^{-1}$ *and* L *is a length scale.*

2.   *For the following find the dimensions of the required variable.*

(a) *Find* $[h]$ *from*

$$-k\frac{\partial u}{\partial x}(L, t) = h\left(u(L, t) - u_e\right)$$

given $[u] = \Theta$, $[u_e] = \Theta$, $[k] = \mathrm{MLT}^{-3}\Theta^{-1}$ *and* $[x] = \mathrm{L}$.

(b) Find $[\alpha]$ from

$$\frac{\partial u}{\partial t} = \alpha \frac{\partial^2 u}{\partial x^2}$$

given $[u] = \Theta$, $[t] = \mathbf{T}$ and $[x] = \mathbf{L}$.

3.   Consider the equation

$$s = s_0 + v_0 t - \frac{1}{2}gt^2.$$

Here s is the position (measured vertically from a fixed reference point) of a body at time t, s_0 is the position at time $t = 0$ and g is the acceleration due to gravity. Is this equation dimensionally correct?

4.   Determine whether the equation

$$\frac{dE}{dt} = \left[mr^2 \left(\frac{d^2\theta}{dt^2} \right) + mgr\sin(\theta) \right] \frac{d\theta}{dt}$$

for the time rate change of total energy E in a pendulum system is dimensionally correct.

5.   Consider a solid metal rod with cross-sectional area A. Suppose that the heat flow is along the axis of the rod and that no heat can escape from the surface of the rod. Further, suppose that due to an electric current, heat is produced at a rate Q joules per unit time per unit volume within the rod. By considering a thin section of thickness δx derive the heat equation

$$\rho c \frac{\partial u}{\partial t} = k \frac{\partial^2 u}{\partial x^2} + Q$$

where ρ, c, and k are the (constant) density, specific heat and conductivity of the material. (This requires a slight modification of the argument used in the notes to derive the one-dimensional heat equation.)

6.   Suppose that in a hollow pipe (cross-section A) a fluid is moving with constant velocity V. Within the fluid heat is being conducted according to Fourier's law with conductivity constant k. The velocity of the fluid and direction of heat conduction may be assumed to be in the same direction as the x-axis which coincides with the axis of symmetry of the pipe.

Consider a section of the pipe between x and $x + \delta x$. Given the temperature $U(x,t)$ derive the advection-conduction equation

$$\rho c \left(\frac{\partial U}{\partial t} + V \frac{\partial U}{\partial x} \right) = k \frac{\partial^2 U}{\partial x^2}.$$

[Hint: This is a modification of the argument that is used to derive the one-dimensional heat conduction equation. There will be an additional flux of heat through the section due to the heat energy which is carried along with the fluid.]

7.   Suppose that the conductivity $k(u)$ is a function of temperature u. By modifying the derivation of the one-dimensional heat equation in the notes obtain the **nonlinear** heat conduction equation

$$\rho c \frac{\partial u}{\partial t} = \frac{\partial}{\partial x}\left(k(u)\frac{\partial u}{\partial x}\right).$$

8.   A solute is diffusing in a mixture in three spatial dimensions. A mass balance equation for the solute is

$$\frac{\partial}{\partial t}\int_V C(\mathbf{x}, t)\, dV = -\oint_S \mathbf{J}(\mathbf{x}, t)\cdot \mathbf{n}\, dS.$$

(a) Describe what each of the two terms in the equation mean physically.

(b) Use Fick's law and the divergence theorem to show that the concentration $C(\mathbf{x}, t)$ satisfies the 3-D diffusion equation

$$\frac{\partial C}{\partial t} = D\nabla^2 C.$$

9.   By considering an arbitrary volume V formulate the 3-D heat equation

$$\frac{\partial u}{\partial t} = \alpha\nabla^2 u$$

where $u(\mathbf{x}, t)$ is the temperature at a point $\mathbf{x}$ and at time t, and α is a positive constant. [*Hint: See previous question.*]

10.   Consider the **equilibrium** heat conduction problem with conductivity, $k(u)$, depending on the temperature u,

$$\frac{d}{dx}\left(k(u)\frac{du}{dx}\right) = 0, \qquad u(0) = 0, \quad u(1) = 10.$$

The next simplest choice for $k(u)$ after the constant case is to assume that the conductivity depends linearly on temperature, i.e. $k(u) = k_0(1 - bu)$ where b and k_0 are positive constants. This corresponds to a substance where the conductivity decreases as the temperature increases.

(a) Show, by solving the equilibrium 1-D heat equation, that

$$u(x) = \frac{1}{b}\left[1 - \sqrt{1 - 2b(10 - 50b)x}\right], \qquad \text{if } b \le \frac{1}{10}.$$

(Make certain discarding the positive sign solution of the quadratic equation is justified.)

(b) Sketch the graph of the equilibrium temperature. Also sketch the constant conductivity case on the same diagram.

(c) Discuss the physical significance of the condition $b \le \frac{1}{10}$.

11.   We consider a problem involving diffusion with a chemical reaction. A gas dissolves into a film of liquid on a surface (see Figure 1.11.1). Inside the film a chemical reaction is going on which causes the gas molecules to disappear thus setting up a concentration gradient which causes more gas molecules to be absorbed into the film. We wish to set up a mathematical model which predicts the rate of absorption of the gas molecules.

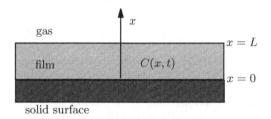

Fig. 1.11.1. Diagram for Question 11. Diffusion with chemical reaction in a thin liquid film on a surface.

Suppose the gas-liquid interface is at $x = L$, where the concentration of gas molecules is always a fixed amount c_0. The bottom of the film is at $x = 0$ where there is no movement of molecules (i.e. the mass flux is zero at $x = 0$). The concentration of gas molecules satisfies the diffusion equation with a negative source term which accounts for the loss of gas molecules due to the chemical reaction. We assume that the rate of loss of molecules is proportional to the concentration. The equation for the concentration is therefore

$$\frac{\partial C}{\partial t} = D\frac{\partial^2 C}{\partial x^2} - \mu C$$

where μ is a positive constant, called the rate constant.

(a) Show, by solving the **appropriate** differential equation, that the **equilibrium** concentration is given by

$$C(x) = \frac{c_0 \cosh(bx)}{\cosh(bL)} \qquad \text{where} \quad b = \sqrt{\mu/D}.$$

(b) Sketch the equilibrium concentration against x.

(c) Calculate the equilibrium mass flux at the liquid-gas interface. Does this have the correct sign?

12.   Consider a rectangular cooling fin, as shown in Figure 1.11.2. Its function is to increase the transport of heat from a wall to the air. We wish to calculate the equilibrium temperature distribution along x and the heat lost to the environment.

The end of the fin is at the ambient air temperature, which we take as $0\,^\circ\text{C}$. Heat flows along the fin from the wall to the air. Heat is also lost from the top and bottom surfaces of the strip according to Newton's law of cooling (heat

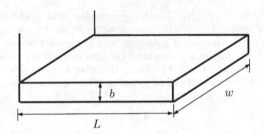

Fig. 1.11.2. A rectangular heat fin. See Question 12 and Question 13.

flux from the surface is proportional to the temperature difference). Assume that the heat loss from the end and the sides is negligible.

(a) Using a one-dimensional approach consider a thin section of thickness δx at x. Let u be the average cross-sectional temperature in the rod at a distance x from the wall. By accounting for the change of heat in the section in a time δt derive the equation for the **equilibrium** temperature

$$\frac{d^2 u}{dx^2} = \frac{2h}{kb} u.$$

(b) The temperature at $x = 0$ is u_0. The boundary condition at $x = L$ is obtained from the above assumptions. Solve this boundary value problem and obtain an expression for the equilibrium temperature as a function of x.

(c) Calculate the total rate of heat lost from the top surface.

13.   This is an alternative derivation for the differential equation in Question 12. Suppose $v(x, y)$ is the two-dimensional equilibrium temperature. Define

$$u(x) = \frac{1}{b} \int_{-b/2}^{b/2} v(x, y)\, dy$$

as the average equilibrium temperature over a cross-section. Integrate the equilibrium heat equation

$$\frac{\partial^2 v}{\partial x^2} + \frac{\partial^2 v}{\partial y^2} = 0$$

and obtain the ordinary differential equation in the previous question.

14.   A thin wire (radius a) is extruded at fixed velocity V_0 through a die (see Figure 1.11.3) and the wire temperature at the die is a fixed value u_0. The wire then passes through the air having temperature u_a for some distance before it is rolled onto large spools at a large distance from the die nozzle.

We wish to investigate the relationship between the wire velocity and the distance between the extrusion nozzle and roll for specific values of V_0.

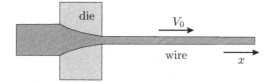

Fig. 1.11.3. Diagram for Question 14. A thin wire being extruded through a die.

(a) Show the temperature $u(x,t)$ in the wire satisfies the partial differential equation

$$\left(\frac{\partial u}{\partial t} + V_0 \frac{\partial u}{\partial x}\right) = \alpha \frac{\partial^2 u}{\partial x^2} - \frac{2h}{a}(u - u_a).$$

Define any additional symbols introduced. [Hint: Neglect any radial heat flow. Consider a section between x and $x + \delta x$. The heat balance then includes heat lost to the surroundings at temperature u_a by Newton's law of cooling.]

(b) For our model let us consider the semi-infinite wire $0 < x < \infty$ and attempt to obtain the **equilibrium** temperature $u(x)$. State the boundary conditions and solve the **appropriate** ordinary differential equation to obtain

$$\frac{u(x) - u_a}{u_0 - u_a} = \exp(\sigma x), \qquad \sigma = \frac{V_0}{2\alpha} - \sqrt{\frac{V_0^2}{4\alpha^2} + \frac{2h}{ka}}$$

(You must be careful to justify the choice of the minus sign when solving the quadratic equation.)

(c) Consider the limiting cases $V_0 \to 0$ and $V_0 \to \infty$.

15. Consider a long nuclear fuel rod which is surrounded by an annular layer of aluminium "cladding" as shown in Figure 1.11.4. Within the fuel rod heat is being produced by fission; this heat source is dependent on position with a source strength varying approximately as

$$Q(r) = Q_0 \left[1 + b\left(\frac{r}{R}\right)^2\right]$$

where r is the radial coordinate and Q_0 and b are constants which are measured by experiment.

In cylindrical polar coordinates, the differential equation for the equilibrium temperature in the fuel rod is

$$k\frac{1}{r}\frac{\partial}{\partial r}\left(r\frac{\partial u_1}{\partial r}\right) + Q(r) = 0$$

where k is the conductivity.

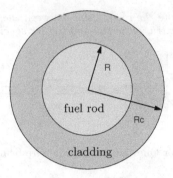

Fig. 1.11.4. Diagram for Question 15. A nuclear fuel rod surrounded by cladding.

(a) *Sketch $Q(r)$.*

(b) *What is the differential equation for the equilibrium temperature $u_2(r)$ in the cladding? What is the general solution of this differential equation?*

(c) *Assuming that the outer surface of the cladding, $r = R_c$, is always at the temperature u_c and the conductivities of the fuel rod and the cladding are k and k_c respectively, show that the equilibrium temperature of the inner surface of the cladding is given by the expression*

$$u_c + \frac{Q_0 R^2}{2k_c}\left(1 + \frac{b}{2}\right)\log\left(\frac{R_c}{R}\right).$$

(d) *Where does the maximum temperature in the fuel rod occur?*

16. *Consider the heat conduction equation*

$$\rho c \frac{\partial u}{\partial t} = k\frac{\partial^2 u}{\partial x^2} + qu$$

with a volumetric source term $Q = qu$ (q is a constant), where the source term is proportional to the temperature $u(x,t)$. Show that the change of dependent variable

$$u(x,t) = v(x,t)e^{qt/\rho c}$$

results in a heat equation for $v(x,t)$ without a source term.

17. *The heat conduction equation in spherical coordinates (with spherical symmetry) is*

$$\frac{\partial u}{\partial t} = \alpha \frac{1}{r^2}\frac{\partial}{\partial r}\left(r^2\frac{\partial u}{\partial r}\right)$$

where α is the (constant) heat diffusivity. Show that the substitution for the dependent variable,

$$v(r,t) = ru(r,t),$$

reduces this equation to the standard cartesian form of the heat/diffusion equation

$$\frac{\partial v}{\partial t} = \alpha \frac{\partial^2 v}{\partial r^2}.$$

18. Consider the transformation of the independent variable x,

$$\xi = x - Vt \qquad \tau = t.$$

(a) Show, that under this transformation, the one-dimensional advection-diffusion equation (with V a constant)

$$\frac{\partial C}{\partial t} + V\frac{\partial C}{\partial x} = D\frac{\partial^2 C}{\partial x^2}$$

transforms to the one-dimensional diffusion equation

$$\frac{\partial C}{\partial \tau} = D\frac{\partial^2 C}{\partial \xi^2}.$$

(This is simply a change of reference frame.)

(b) Suppose that V is a function of time t. What should the transformation be in this case in order to get the same equation?

19. The equations describing a convection-diffusion problem in heat transport are

$$\frac{\partial u}{\partial t} + v\frac{\partial u}{\partial x} = \alpha\frac{\partial^2 u}{\partial x^2}, \qquad u(0,t) = u_0, \quad u(x,0) = 0, \quad u(\ell,t) = u_1,$$

where $u(x,t)$ is the temperature, v is the (constant) velocity and α is the thermal diffusivity. Consider the following scaling,

$$X = \frac{x}{\ell}, \quad T = \frac{\alpha t}{\ell^2}, \quad U = \frac{u}{u_1}, \quad V = \frac{\ell v}{\alpha}.$$

Show that the above equations may be expressed in the dimensionless form

$$\frac{\partial U}{\partial T} + \epsilon_1 \frac{\partial U}{\partial X} = \frac{\partial^2 U}{\partial X^2}, \qquad U(0,T) = \epsilon_2 \quad U(X,0) = 0, \quad U(1,T) = 1,$$

giving formulae for the dimensionless constants ϵ_1 and ϵ_2.

20.   A missile, which has mass m, is projected upwards from the Earth's surface with initial speed v_0. The missile is subject to a gravitational field which varies with the inverse square distance from the Earth's centre. Let a denote the radius of the Earth, let g denote the gravitational acceleration at the Earth's surface, and let $r(t)$ be the radial distance of the missile from the Earth's surface at time t. If we neglect air resistance the governing equations of motion of the missile are

$$\frac{d^2 r}{dt^2} = -\frac{ga^2}{(r+a)^2}, \qquad r(0) = 0, \qquad \frac{dr}{dt}(0) = v_0.$$

(a) Consider the following scales: v_0^2/g for r and v_0/g for t. We define the new variables

$$R = \frac{r}{v_0^2/g}, \qquad T = \frac{t}{v_0/g}.$$

Check that R and T are dimensionless.

(b) Substitute these into the equations of motion and obtain

$$\frac{d^2 R}{dT^2} = -\frac{1}{(\epsilon R + 1)^2}, \qquad R(0) = 0, \qquad \frac{dR}{dT}(0) = 1.$$

Give the parameter ϵ.

(c) How do the equations simplify for $\epsilon \ll 1$? What do the simplified equations correspond to physically?

(d) Suppose we use a different set of scales for the previous problem:

$$R = r/a, \qquad T = v_0 t/a.$$

Write down the dimensionless differential equation and initial conditions. Explain why this scaling is inappropriate for small values of $V^2/(ga)$. (You can give a mathematical and a physical reason here.)

21.   A pendulum is executing small vibrations.

(a) Use dimensional analysis to show that it is impossible on purely dimensional grounds for the period τ to depend **only** on the length ℓ of the pendulum and the mass m of the bob.

(b) Suppose you now assume the period also depends on g the acceleration due to gravity. What does dimensional analysis say about this?

22.   Consider the (incompressible) flow of a liquid in a circular pipe. Assume the pressure difference between the two ends of the pipe, ΔP, depends only on the pipe length ℓ, radius R, the maximum speed of the fluid U, the fluid viscosity μ (where $[\mu] = \text{ML}^{-1}\text{T}^{-1}$) and density ρ. Use dimensional analysis to express ΔP as a function of dimensionless variables.

One answer is

$$\frac{\Delta P}{\rho U^2} = F\left(\frac{\rho U R}{\mu}, \frac{\rho U \ell}{\mu}\right)$$

but there are others.

23. A windmill is being rotated by air flow to produce power to pump water. It is desired to find the power output P of the windmill. Assume that P is a function of the density of air ρ, the viscosity of air μ, diameter of the windmill d, wind speed v, and the rotational speed ω (measured in radians per second).

(a) Argue why $[\omega] = \mathbf{T}^{-1}$.

(b) Using dimensional analysis, find a dimensionless relationship for P.

24. The thrust (force) T developed by a ship's propeller in deep water depends on the radius a of the propeller, the number of revolutions per minute n, the velocity V with which the ship advances, the gravitational constant g and the density ρ of the water.

Show that

$$\frac{T}{\rho a^2 V^2} = \Phi\left(\frac{an}{V}, \frac{ag}{V^2}\right)$$

for some function Φ of two variables. [Hint: First argue $[n] = \mathbf{T}^{-1}$.]

25. Many cookbooks say that a roast should be cooked for x minutes per kilogram. Carry out the following dimensional analysis. Assume that (for a given cooking temperature) the time τ for the centre of the roast to reach the required temperature depends only on the mass of the roast m, the density of the meat ρ and the thermal diffusivity of the meat α.

(a) Show that

$$\tau \quad \text{is proportional to} \quad m^{2/3}.$$

(b) Discuss whether the cookbooks are correct in assuming that the roast should be cooked for a certain number of minutes per kilogram.

26. Use dimensional analysis to predict the crater volume V produced by an explosion on the surface of the earth. Assume that the size of the crater depends on the mass of the explosive and the density of the soil. Hence explain whether a higher soil density will produce a larger or smaller crater.

27.   The lift force F on a missile depends on the length of the missile ℓ, its velocity v and diameter d; it also depends on the air density ρ, viscosity μ, gravity g, and the speed of sound in air, c.

Show that

$$F = \rho v^2 \ell^2\, \hat{f}\left(\frac{d}{r}, \frac{\mu}{\rho v r}, \frac{c}{v}, \frac{\ell g}{v^2}\right)$$

where $\hat{f}$ is some function of four variables.

28.   Suppose (p_1, p_2) and (q_1, q_2) denote two sets of measurements of primary quantities (i.e. quantities whose dimensions are fundamental dimensions). Let $f(p_1, p_2)$ and $f(q_1, q_2)$ be two measurements of the same secondary quantity. We are going to show that $f(p_1, p_2) = C p_1^{a_1} p_2^{a_2}$.

(a) Give reasons why we can postulate

$$\frac{f(x_1 p_1, x_2 p_2)}{f(p_1, p_2)} = \frac{f(x_1 q_1, x_2 q_2)}{f(q_1, q_2)}. \qquad (*)$$

(b) By taking $\dfrac{\partial}{\partial x_1}$ of $(*)$ and then setting $x_1 = 1$, deduce that

$$f(p_1, p_2) = g(p_2) p_1^{a_1},$$

where a_1 is a constant (with respect to p_1 and p_2) and g is an arbitrary function.

(c) Hence deduce that

$$f(p_1, p_2) = C p_1^{a_1} p_2^{a_2}$$

where C is a constant (with respect to p_1 and p_2).

This result may be extended to $f(p_1, p_2, p_3, \ldots, p_n)$ for any n. (See Lin and Segel (1974)).

2

Case Study: Continuous Casting

The case study in this chapter involves finding the puddle length in a continuous casting operation. This involves calculating how fast molten steel solidifies. We introduce the Boltzmann similarity transformation as a way of solving the 1-D heat equation with a moving boundary (the boundary between molten and solidified steel, which changes with time). This technique reduces the PDE into an ordinary differential equation (ODE) and a parameter describing the moving boundary position is obtained as the solution of a transcendental equation.

2.1 Introduction to the case study problem

Of great interest in many industrial applications are those problems which involve a change of phase — from solid to liquid or liquid to solid, for example. These are also interesting mathematically because few exact solutions are known. Hence the ones for which analytic solutions are known give considerable insight into the physical processes involved. We develop a mathematical model to examine the feasibility of casting steel sheets by pouring molten metal onto a cooled rotating drum. This model involves the concept of a moving solidification boundary. The problem was brought to the 1985 Australian Mathematics in Industry Workshop by the Research Laboratories of BHP in Melbourne. It was reported in Barton (1985).

Background

A conventional method of producing steel sheets involves rolling steel billets down to the required thickness. This can be costly in both resources and time. Also it is difficult to produce very long sheets of steel by this method. One alternative is to pour molten steel onto a rotating drum which is cooled by water flowing through it (see Figure 2.1.1). Provided the molten steel solidifies quickly enough then the steel sheet can be removed from the drum. The Australian steel manufacturing company BHP were interested in using this method to cast sheet steel between 1 mm and 10 mm thick. The drum speed was to be about $1\,\mathrm{m\,s^{-1}}$.

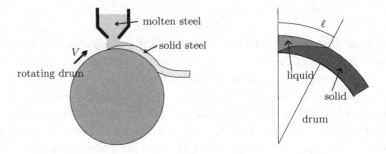

Fig. 2.1.1. Schematic diagrams of continuous casting on a water cooled drum.

The thickness of the steel sheet is controlled by varying the flow rate of molten steel from the container. Thus to make the sheet thicker you would increase the flow rate of the molten steel onto the drum.

As the molten steel is poured onto the drum it begins to solidify as heat passes from the molten steel into the water cooled drum. In doing so a "puddle" of molten steel exists above the solidified steel. For this process to work effectively we do not want the liquid metal to run off as the drum rotates. Thus the molten steel must solidify *before* it gets too far around the drum.

We need to undertake a feasibility study of the solidification process. To do this we formulate a mathematical model of the important physical aspects of the problem. The model does not have to be an exact representation of the problem, however, we hope to use it to perform calculations that are accurate at least to an order of magnitude. Further refinements of the model (with possible computer calculations) would then be justified if the initial calculations indicated whether the process was feasible.

A mathematical model

The aim of our mathematical model, then, is to determine the "puddle length". This is the distance from the point where the molten steel is poured onto the drum (and begins to solidify) to the point where the whole layer has solidified. For a drum whose surface rotates with speed V, the puddle length ℓ is simply given by

$$\ell = Vt_h \tag{1}$$

where t_h is the time for the metal to solidify through the full thickness h of the metal sheet, and V is the speed of the surface of the rotating drum.

Thus what we really need to calculate is the quantity t_h, the time taken for the molten steel to solidify a distance h. We try to do that by modelling the heat transfer and solidification of the molten metal by a one-dimensional approach. We first observe that if we fix our frame of reference (co-ordinate system) on the surface of the drum then we have a moving boundary which moves from the surface of the drum to the liquid-air interface, as in Figure 2.1.2.

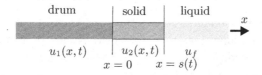

Fig. 2.1.2. 1-D model for the solidification and heat transport processes in the molten steel and water cooled drum.

The temperature of the molten metal is very close to the solidification temperature, $u_f \simeq 1400\,^\circ\mathrm{C}$ for steel. We will neglect radiative heat transfer† from the liquid-air interface, assuming that all the heat flows towards the cooled drum. Thus we can assume that the temperature in the liquid phase is a constant, u_f. We take the temperature in the

---

† Neglecting radiative transfer, for this problem, can be justified by a simple model of one-dimensional heat conduction with radiative transfer from the liquid-air interface and ignoring solidification. The calculations in Barton (1985) showed the radiative transfer term was small and that the solidification occurs from the drum.

solidified metal, u_2, and the temperature in the cooled rotating drum, u_1, to be functions of x and t.

The radius of the drum is large compared to the thickness of the metal. Thus, in our one-dimensional model we take the drum to be the semi-infinite region $-\infty < x < 0$. The temperature of the drum at a large distance from the surface is $U \simeq 150\,°C$.

Governing equations

Let us define $u_2(x,t)$ to be the temperature of the solidified metal at time t and at distance x from the drum surface. Let us also denote the position of the moving interface between the liquid and the solid to be at $x = s(t)$, for some function s to be determined. We also let $u_1(x,t)$ denote the temperature in the drum. Our aim is to solve for $u_1(x,t)$, $u_2(x,t)$ and $s(t)$.

Both u_1 and u_2 satisfy the 1-D heat conduction equations

$$\frac{\partial u_1}{\partial t} = \alpha_1 \frac{\partial^2 u_1}{\partial x^2},$$
$$\frac{\partial u_2}{\partial t} = \alpha_2 \frac{\partial^2 u_2}{\partial x^2}. \tag{2}$$

Note that each equation has a different thermal diffusivity α_1 and α_2 since the solidified steel and the copper drum are different materials.

We require two boundary conditions for each of the above partial differential equations since they are both second-order in x. We need one boundary condition at $x = -\infty$, two at $x = 0$ and one at the moving boundary between solid and liquid.

The temperature at the core of the copper drum was provided as u_d. At $x = 0$ the temperature and heat flux must be continuous. On the solid-liquid boundary the temperature must also be continuous. We shall let $x = s(t)$ denote the position of this moving boundary. So $x < s(t)$ denotes the solid steel and $x > s(t)$ denotes the molten steel. We let u_d denote the temperature of the core of the copper drum, and u_f the solidifying (freezing) temperature of molten steel. The boundary conditions are

$$u_1(-\infty, t) = u_d,$$
$$u_1(0, t) = u_2(0, t),$$
$$-k_1 \frac{\partial u_1}{\partial x}(0, t) = -k_2 \frac{\partial u_2}{\partial x}(0, t), \tag{3}$$
$$u_2(s(t), t) = u_f,$$

where k_1 and k_2 are the conductivities of copper and solid steel.

Although, in principle, we can solve for both u_1 and u_2, we need an extra boundary condition. This will let us also solve for the position of the moving boundary $s(t)$. This extra boundary condition comes from considering the **latent heat** released when the molten steel solidifies. We shall determine this boundary condition (known as a Stefan condition) after investigating some basic physical concepts regarding latent heat.

Latent heat

At certain temperatures, heat energy added to a material can alter the physical structure of the material. For example, at $0\,^\circ$C with additional heat ice breaks its intermolecular bonds and changes into liquid water if a sufficient amount of heat is added to it. Similarly, if we remove heat from water at $0\,^\circ$C then the water freezes and becomes ice. This is called a **phase change**.

The amount of heat required to do this is called the **latent heat**. More specifically, the **latent heat of fusion** for a given material is the amount of heat required to convert a mass of solid, at its melting temperature, into a liquid at the same temperature. In the reverse process of solidification, the latent heat of fusion is the amount of heat released when molten material solidifies.

The latent heat for a phase change is usually expressed per unit mass, and here we give it the symbol λ. We call this quantity the **specific latent heat**. The SI units are joules kg^{-1}. Some typical values of λ are given in Table 2.1.1.

A similar description can be given for the process of boiling a liquid, or the reverse process of condensation. Here, we refer to the specific latent heat of **vaporisation**. As a comparison, some values for specific latent heats for vaporisation of liquids are given in Table 2.1.2.

Table 2.1.1. *Melting temperature u_f and specific latent heat of fusion λ for water and various metals. Source: Science Data Handbook (1971).*

	u_f °C	$\lambda\,\mathrm{J\,kg^{-1}}$
Ice	0	33×10^4
Solder	217	190×10^4
Lead	327	2.6×10^4
Aluminium	659	38×10^4
Copper	1 083	21×10^4
Gold	1 067	7.0×10^4
Iron	1 537	27×10^4
Steel	1 440	27×10^4

Table 2.1.2. *Boiling temperature u_v and specific latent heat of vaporisation λ.*

	u_v °C	$\lambda\,\mathrm{J\,kg^{-1}}$
Water	100	2.26
Ethyl alcohol	79	8.5×10^{-1}
Helium	−269	2.5×10^{-2}

The Stefan condition

We now derive the boundary condition on the moving boundary $x = s(t)$. To do this we assume the moving boundary advances a distance δs in a time δt. This is illustrated in Figure 2.1.3.

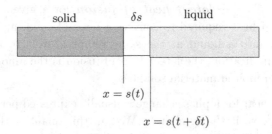

Fig. 2.1.3. The solid-liquid interface advancing a distance δs in time δt.

For the region of thickness δs, we equate the heat released by solidification (latent heat) to the amount of heat removed by conduction. The following example shows how to use conservation of energy (heat balance) to obtain an additional boundary condition at $x = s(t)$.

Example 1:   *Obtain the boundary condition representing the absorption of latent heat.*

Solution: *A word equation which expresses conservation of heat for the region $s(t)$ to $s(t + \delta t)$ is*

$$\left\{ \begin{array}{c} \text{latent heat} \\ \text{liberated by} \\ \text{solidifying mass} \end{array} \right\} = \left\{ \begin{array}{c} \text{heat conducted} \\ \text{back through} \\ \text{cross-section at } x = s(t) \end{array} \right\}.$$

Let λ be the specific latent heat (per unit mass) of the metal. The total mass of the material which has solidified in the given time interval is $\rho A \delta s = \rho A(s(t + \delta t) - s(t))$. Here ρ is the density of the material (assumed the same, here, for solid and liquid phases) and A is some cross-sectional area. Thus

$$\left\{ \begin{array}{c} \text{latent heat} \\ \text{liberated by} \\ \text{solidifying mass} \end{array} \right\} = \lambda \rho A \delta s. \tag{4}$$

Since the liquid metal is assumed to be at a uniform temperature there is no temperature gradient in the direction of the liquid metal. All the heat flows through the boundary $x = s(t)$ back into the solid, towards the drum surface. The amount of heat conducted through $x = s(t)$ is obtained by multiplying the heat flux by the cross-sectional area A and the time interval δt. Using Fourier's law, $J = -k\partial u/\partial x$ is the heat flux in the positive x-direction. Thus, we have

$$\left\{ \begin{array}{c} \text{heat conducted} \\ \text{back through} \\ \text{cross section at } x = s(t) \end{array} \right\} = -JA\delta t = k \left. \frac{\partial u}{\partial x} \right|_{x=s(t)} A\delta t. \tag{5}$$

(The minus sign indicates that heat is conducted in the negative x-direction, towards the solid.)

We now equate (4) and (5). We divide through by $A\delta t$ and then let $\delta t \to 0$ (corresponding to $\delta s \to 0$). We thus obtain, in the limit,

$$k\frac{\partial u}{\partial x}(s(t), t) = \lambda \rho \frac{ds}{dt}. \tag{6}$$

This boundary condition (6) is known as the **Stefan condition** after J. Stefan who derived it around the turn of the 19th century to study the freezing of water in the ground in order to calculate the thickness of the polar ice caps. There is a more complicated version of the Stefan condition when the liquid phase is at a different temperature to the freezing temperature. There is also a Stefan condition for melting problems.

These are explored in the problems at the end of this chapter. Various Stefan-type problems are reviewed in Carslaw and Jaeger (1959).

Looking at some simpler problems

We now need to develop the techniques to solve the partial differential equations (2) with the boundary conditions (3) and the Stefan condition (6). This will give us the time for the moving boundary to solidify a distance h and hence we can calculate the puddle length ℓ using (1).

In the next two sections we shall look at two simpler problems. The first problem is that of heat conduction in a semi-infinite region, where we introduce the approach of using a similarity solution. In the following section, we look at a simpler phase-change problem, using the same approach. Finally, we shall solve the equations formulated here and use the solution to estimate the puddle length.

2.2 The Boltzmann similarity solution

One useful method for finding solutions of partial differential equations is to use a transformation that changes the governing equations into a different form that is easier to handle. This leads to Boltzmann's similarity solution of the heat equation.

An example problem

Let us consider the problem of heat conduction in a semi-infinite region $0 < x < \infty$. Suppose the region is initially at zero temperature and the end of the region is suddenly set and maintained at some constant temperature u_1. We wish to determine how the temperature changes with time at various points along the x-axis.

The governing partial differential equation is the standard 1-D heat equation

$$\frac{\partial u}{\partial t} = \alpha \frac{\partial^2 u}{\partial x^2} \tag{1}$$

The initial condition and two boundary conditions are

$$u(x,0) = 0, \qquad (2)$$

$$u(0,t) = u_1, \qquad (3)$$

$$u(\infty,t) = 0. \qquad (4)$$

The Boltzmann transformation

Suppose we wanted to express the solution of this problem in dimensionless variables. The combination u/u_1 is dimensionless. In this problem there is no natural length scale (as there would be if we were solving for a finite region of length ℓ). Given the parameter α has dimensions L^2T^{-1} then only the combination $\sqrt{\alpha t}$ has dimensions of length. This suggests the dimensionless temperature can only be a function of $x/\sqrt{\alpha t}$ if we are to be able to express the solution in dimensionless terms. This situation has a tremendous advantage for solving the partial differential equation, since it suggests the solution can only depend on one variable, $u = f(x/\sqrt{\alpha t})$. You might like to go back and read the section on dimensional analysis (Section 1.10) to show this formally.

Let us write

$$u(x,t) = f(\eta), \qquad \text{where} \quad \eta = \frac{x}{\sqrt{\alpha t}}. \qquad (5)$$

where f is some function of **one** variable. This transfomation is known as the **Boltzmann transformation** and η is the **Boltzmann similarity variable**.

Because the temperature is a function of the single similarity variable $x/\sqrt{\alpha t}$, we aim to reduce the partial differential equation (1) and boundary conditions into an ordinary differential equation for f, with matching boundary conditions for f.

Reduction of variables

We now substitute the Boltzmann transformation (5) into the partial differential equation (1) to obtain an ordinary differential equation. This involves using the chain rule. The following example shows how to do this.

Example 1: Use the transformation

$$u(x,t) = f(\eta), \qquad \eta = \frac{x}{\sqrt{\alpha t}}$$

to reduce the partial differential equation (1) to an ordinary differential equation.

Solution: For $\partial u / \partial t$ we obtain, using the chain rule,

$$\frac{\partial u}{\partial t} = \frac{df}{d\eta} \frac{\partial \eta}{\partial t}.$$

Since $\eta = \alpha^{-1/2} x t^{-1/2}$ then

$$\frac{\partial u}{\partial t} = -\frac{1}{2} x \alpha^{-1/2} t^{-3/2} \frac{df}{d\eta}.$$

We can now eliminate x using $x = \alpha^{1/2} t^{1/2}$ to obtain

$$\frac{\partial u}{\partial t} = -\frac{1}{2} t^{-1} \eta \frac{df}{d\eta}. \tag{6}$$

First, let us consider the first derivative with respect to x. We obtain

$$\frac{\partial u}{\partial x} = \frac{df}{d\eta} \frac{\partial \eta}{\partial x} = \alpha^{-1/2} t^{-1/2} \frac{df}{d\eta}.$$

For the second derivative we must use the chain rule twice,

$$\frac{\partial^2 u}{\partial x^2} = \frac{\partial}{\partial x}\left(\frac{\partial u}{\partial x}\right) = \frac{d}{d\eta}\left(\frac{df}{d\eta} \frac{\partial \eta}{\partial x}\right) \times \frac{\partial \eta}{\partial x}.$$

Substituting $\eta = \alpha^{-1/2} x t^{-1/2}$ we obtain

$$\frac{\partial^2 u}{\partial x^2} = \alpha^{-1/2} t^{-1/2} \frac{d}{d\eta}\left(\frac{df}{d\eta}\right) \frac{\partial \eta}{\partial x} = \alpha^{-1/2} t^{-1/2} \frac{d^2 f}{d\eta^2} \alpha^{-1/2} t^{-1/2}.$$

This simplifies to

$$\frac{\partial^2 u}{\partial x^2} = \alpha^{-1} t^{-1} \frac{d^2 f}{d\eta^2}. \tag{7}$$

We now substitute the expressions for each derivative, (6) and (7) back into the partial differential equation (1). We thus obtain

$$-\frac{1}{2} t^{-1} \eta \frac{df}{d\eta} = \alpha \left(\alpha^{-1} t^{-1} \frac{d^2 f}{d\eta^2}\right).$$

Dividing through by t^{-1} we now obtain the **ordinary** differential equation

$$-\frac{1}{2}\eta\frac{df}{d\eta} = \frac{d^2f}{d\eta^2}. \qquad (8)$$

Note that there is no x or t left in the differential equation. The differential equation is only in terms of the single similarity variable η. This is what we wanted to obtain — a reduction of the partial differential equation to an ordinary differential equation.

Solving the ordinary differential equation

We can solve this differential equation by recognising that the second-order differential equation in f is also a first-order differential equation in f' since f does not appear explicitly. The following example shows how to solve this ordinary differential equation.

Example 2: *Find the general solution of the ordinary differential equation (ODE) (8).*

Solution: *Writing f' for $df/d\eta$ the ODE (8) becomes*

$$\frac{df'}{d\eta} = -\frac{1}{2}\eta f'$$

which is a first-order separable differential equation. Solving this we obtain a general solution

$$f'(\eta) = c_1 e^{-\eta^2/4}$$

where c_1 is a constant of integration.

Integrating f' to obtain f, we obtain

$$f(\eta) = c_1 \int_0^\eta e^{-\eta_1^2/4}\,d\eta_1 + c_2$$

where c_2 is another integration constant. Letting $v = \eta_1/\sqrt{4}$ we can write this as

$$f(\eta) = 2c_1 \int_0^{\eta/2} e^{-v^2}\,dv + c_2. \qquad (9)$$

The error function

Unfortunately we cannot evaluate the integral in (9) in terms of elementary functions. However, an integral similar to that in (9) frequently occurs in mathematical applications, especially statistics and probability. This integral defines a function, known as the **error function**, denoted by erf, and given by

$$\text{erf}(z) = \frac{2}{\sqrt{\pi}} \int_0^z e^{-v^2} \, dv. \qquad (10)$$

Our aim is to express the solution (9) in terms of the error function since this gives a simpler form of the solution that is quite useful once the error function and its properties are familiar. The general solution (9) may thus be written as

$$f(\eta) = C_1 \, \text{erf}(\eta/2) + C_2 \qquad (11)$$

where $C_1 = \sqrt{\pi}c_1$ and $C_2 = c_2$ are a convenient relabelling of the arbitrary constants.

We now explore briefly some properties of the error function, defined by equation (10). Some useful properties of the error function are as follows:

- $\text{erf}(0) = 0$.
- $\text{erf}(\infty) = 1$.
- $\dfrac{d}{dz} \text{erf}(z) = \dfrac{2}{\sqrt{\pi}} e^{-z^2}$.
- erf is monotonic increasing.
- $\text{erf}(-z) = -\text{erf}(z)$, (i.e. erf is antisymmetric).

These are all very easy to prove (with the exception of the second property, which requires multiple integrals). A sketch of the graph of the error function is given below in Figure 2.2.1.

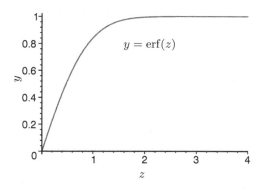

Fig. 2.2.1. Graph of the error function, erf.

Boltzmann similarity solution

Going back to the substitution $u(x,t) = f(\eta)$, where $\eta = x/\sqrt{\alpha t}$, we can write the solution (11) as

$$u(x,t) = C_1 \operatorname{erf}(x/\sqrt{4\alpha t}) + C_2. \qquad (12)$$

This is known as the **Boltzmann similarity solution** of the heat equation. It is applicable for problems defined on semi-infinite regions (which do not have a natural length scale).

Applying boundary conditions

We now solve the problem described by the partial differential equation (1) with the initial and boundary conditions (2), (3) and (4). We start with the Boltzmann similarity solution (12), which is a two-parameter solution of the heat equation. The following example shows how to do this.

Example 3: *Using the Boltzmann similarity solution*

$$u(x,t) = C_1 \operatorname{erf}(x/\sqrt{4\alpha t}) + C_2 \qquad (13)$$

apply the boundary conditions (3) and (4)

$$u(0,t) = u_1, \qquad u(\infty,t) = 0.$$

Solution: *Using the boundary condition $u(0,t) = u_1$, when $x = 0$, then we obtain*

$$C_1 \, \mathrm{erf}(0) + C_2 = u_1.$$

Since $\mathrm{erf}(0) = 0$, then we obtain $C_2 = u_1$.

Similarly, $u(\infty, t) = 0$ gives the equation

$$C_1 \, \mathrm{erf}(\infty) + C_2 = 0.$$

Since $\mathrm{erf}(\infty) = 1$, and substituting $C_2 = u_1$, we obtain $C_1 + u_1 = 0$. The solution (13) is therefore

$$u(x,t) = u_1 \left\{ 1 - \mathrm{erf}\left(\frac{x}{\sqrt{4\alpha t}} \right) \right\}. \tag{14}$$

Note that we could have used the initial condition (2) instead of the boundary condition (4), and would have come up with the same equation (since $t = 0$ gives $\mathrm{erf}(\infty)$, just as $x = \infty$). This is because the initial condition is linked to the boundary condition at infinity through the Boltzmann similarity variable. Physically, this means that the point $x = \infty$ remains at the initial temperature in finite time. In terms of the similarity variables the boundary condition (3) becomes $f(0) = 0$ and the boundary condition (2) and the initial condition (2) both become $f(\infty) = 0$.

A sketch is given below in Figure 2.2.2 of the temperature distribution for different times, for $u_1 = 600\,^\circ\mathrm{C}$ and where $\alpha = 2 \times 10^{-4}$. The temperature distribution is slowly moving towards a uniform distribution.

Applicability of Boltzmann's similarity solution

The Boltzmann transformation will not work for all problems involving the heat equation. It is usually restricted to problems of infinite or semi-infinite domains. It won't work for problems involving finite domains, since the width of the domain introduces a natural length scale into the problem, which then means that the temperature does not have to be only a function of the combination $x/\sqrt{\alpha t}$. An exception is the moving boundary problem of Section 2.3, where the region is finite, but the length is not fixed — it changes as solidification occurs. The following example illustrates this.

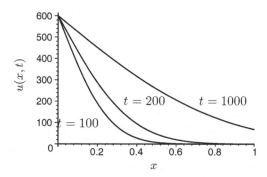

Fig. 2.2.2. Temperature distribution for different times (in seconds) for heat conduction in a semi-infinite region with prescribed temperature $u_1 = 600\,^\circ\mathrm{C}$ at $x = 0$. We have also set $\alpha = 2 \times 10^{-4}$.

Example 4: *Show the Boltzmann similarity solution will not work for the boundary condition $u(1,t) = 100$.*

Solution: *Substituting $x = 1$ into the Boltzmann similarity solution gives*

$$100 = C_1 \operatorname{erf}\left(\frac{1}{\sqrt{4t}}\right) + C_2.$$

Clearly, C_1 and C_2 cannot both be constant if this equation is to be satisfied, so we cannot apply this boundary condition.

It can also be shown that problems that introduce a length scale in some other way also cannot use Boltzmann's similarity solution. An example is the Newton cooling condition where the combination k/h has dimensions of length. Nevertheless, when Boltzmann's similarity solution can be applied, it is a useful method for obtaining a solution of the heat and diffusion equations.

Summary

A two-parameter solution of the heat equation

$$\frac{\partial u}{\partial t} = \alpha \frac{\partial^2 u}{\partial x^2}$$

is the Boltzmann similarity solution

$$u(x,t) = C_1 \operatorname{erf}(x/\sqrt{4\alpha t}) + C_2$$

where C_1 and C_2 are arbitrary constants. This solution is useful for semi-infinite problems, where there is no natural length scale.

2.3 A moving boundary problem

We use the Boltzmann similarity solution to solve a simple moving boundary problem. We see how the Stefan condition is used to determine the position of an advancing solidification front. An alternative approximate technique is also presented — finding the pseudo-steady-state solution.

Example problem

Let us consider a semi-infinite region $0 < x < \infty$ consisting of a material which is solid for $0 < x < s(t)$ and liquid for $x > s(t)$. We assume the initial temperature of the liquid is the freezing temperature, at $0\,^\circ\mathrm{C}$. We also assume the end $x = 0$ is maintained at temperature $-1\,^\circ\mathrm{C}$.

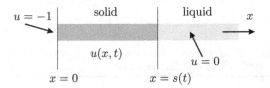

Fig. 2.3.1. One-dimensional model for freezing of a liquid.

Let $u(x,t)$ denote the temperature of the solid phase. The governing partial differential equation is the 1-D heat conduction equation

$$\frac{\partial u}{\partial t} = \frac{\partial^2 u}{\partial x^2} \qquad 0 < x < s(t). \tag{1}$$

We have the boundary conditions

$$u(0,t) = -1, \tag{2}$$

$$u(s(t),t) = 0. \tag{3}$$

Here $x = s(t)$ denotes the (unknown) position of the moving boundary at time t. We also have a Stefan condition

$$\frac{\partial u}{\partial x}(s(t), t) = \frac{ds}{dt}. \tag{4}$$

These equations correspond to the solidification of a material where the units have been chosen so that all the physical constants (e.g. α, ρ, k and λ) have the value 1. (This is equivalent to scaling the variables.) We did not specify an initial condition for $u(x,t)$. This is because there is no solid at $t = 0$, i.e. u does not exist at $t = 0$. However, we do have an initial condition for $s(t)$, namely $s(0) = 0$, which states that the solidification boundary starts from $x = 0$.

Solution using the Boltzmann transformation

We shall see if the Boltzmann similarity solution can be used here. From the previous section, the Boltzmann similarity solution of the heat equation (with $\alpha = 1$) is

$$u(x,t) = C_1 \operatorname{erf}\left(\frac{x}{\sqrt{4t}}\right) + C_2, \tag{5}$$

where C_1 and C_2 are arbitrary constants. (Recall that this was derived using the Boltzmann transformation $u(x,t) = f(\eta)$ with $\eta = x/\sqrt{t}$, which reduced the heat equation to an ordinary differential equation.)

Form of the moving boundary

To solve this problem we need to find **both** the temperature $u(x,t)$ and the moving boundary $s(t)$. The following example shows how to determine the form of $s(t)$ in terms of an unknown proportionality constant. It can be obtained from the boundary condition expressing the temperature at the moving boundary.

Example 1: *Find the form of $s(t)$ using the boundary condition $u(s(t), t) = 0$ and the Boltzmann similarity solution.*

Solution: *On applying boundary condition (3), $u(s(t), t) = 0$, we obtain*

$$C_1 \operatorname{erf}\left(\frac{s(t)}{\sqrt{4t}}\right) + C_2 = 0. \tag{6}$$

Since C_1 and C_2 are constants, the only way this can be satisfied is if $s(t)/\sqrt{4t}$ is a constant. We denote this constant by m. This **requires**

$$s(t) = m\sqrt{4t} \qquad (7)$$

where m is some constant that is yet to be determined.

Once we have a value for m we will know the position of the moving boundary at any time t. We can now find the temperature, in terms of the unknown constant m. This is done in the following example.

Example 2:   *Given the general solution (5) and the form of the moving boundary (7) determine the constants C_1 and C_2.*

Applying the boundary condition $u(0,t) = -1$ to the general solution (5) we obtain $(C_1 \times 0) + C_2 = -1$, since $\mathrm{erf}(0) = 0$. Hence $C_2 = -1$. Applying the boundary condition $u(s(t),t) = 0$, and using (7) we obtain the equation

$$C_1 \, \mathrm{erf}\left(\frac{m\sqrt{4t}}{\sqrt{4t}}\right) + C_2 = 0$$

and hence

$$C_1 \, \mathrm{erf}(m) + C_2 = 0.$$

Since we have already shown $C_2 = -1$ then this determines C_1 as $C_1 = 1/\,\mathrm{erf}(m)$.

Substituting back for C_1 and C_2, the solution for the temperature is

$$u(x,t) = \frac{\mathrm{erf}\left(x/\sqrt{4t}\right)}{\mathrm{erf}(m)} - 1. \qquad (8)$$

This is not yet a complete solution since we still have to determine the unknown constant m.

Applying the Stefan condition

We need to substitute the solution for the temperature, (8), into the Stefan condition (4). The procedure is carried out in the following example.

Example 3:   *Apply the Stefan condition and hence obtain an equation for the constant m.*

Solution: *We first calculate*

$$\frac{\partial u}{\partial x}(x,t) = \frac{1}{\mathrm{erf}(m)}\frac{\partial}{\partial x}\left\{\mathrm{erf}(x/\sqrt{4t})\right\}. \tag{9}$$

From the definition of the error function we obtain

$$\frac{d}{dz}\,\mathrm{erf}(z) = \frac{2}{\sqrt{\pi}}e^{-z^2}.$$

Now using the chain rule we have

$$\frac{\partial}{\partial x}\,\mathrm{erf}(x/\sqrt{4t}) = \frac{2}{\sqrt{\pi}}\times\frac{1}{\sqrt{4t}}e^{-x^2/(4t)}.$$

Hence

$$\frac{\partial u}{\partial x} = \frac{1}{\mathrm{erf}(m)}\frac{2}{\sqrt{\pi}}\times\frac{1}{\sqrt{4t}}e^{-x^2/(4t)}.$$

Evaluating at $x = s(t) = m\sqrt{4t}$, we obtain

$$\frac{\partial u}{\partial x}(s(t),t) = \frac{1}{\mathrm{erf}(m)}\frac{1}{\sqrt{4t}}\frac{2}{\sqrt{\pi}}e^{-m^2}.$$

Also substituting $x = m\sqrt{4t}$ into the RHS of the Stefan condition gives

$$\frac{1}{\mathrm{erf}(m)}\frac{1}{\sqrt{4t}}\frac{2}{\sqrt{\pi}}e^{-m^2} = \frac{1}{2}\sqrt{4}mt^{-1/2}.$$

Note that the $t^{-1/2}$ terms cancel and rearranging the equation, so that the unknown m is on one side, we have

$$m\,\mathrm{erf}(m)e^{m^2} = \frac{1}{\sqrt{\pi}} \tag{10}$$

Equation (10) is a transcendental equation as it cannot be solved in terms of elementary functions. A common transcendental equation that students often meet first is $x = \sin(x)$. Usually, we obtain solutions of transcendental equations graphically (or numerically, using the bisection method or Newton's method, for example). Thus, given the value of σ, we solve (10) for m.

Interpretation of results

To find the root graphically we plot both sides of the equation (10),

$$y = m\,\mathrm{erf}(m)e^{m^2} \qquad \text{and} \qquad y = \pi^{-1/2}.$$

From Figure 2.3.2 the intersection point gives the solution approximately as $m \simeq 0.6$.

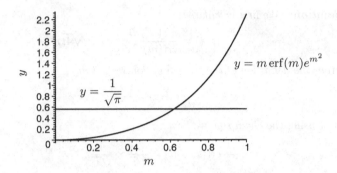

Fig. 2.3.2. Graph showing the root of the transcendental equation.

We can also get a more accurate value by finding a numerical solution (for example, using Newton's method). This is easily done using the software package Maple, with the Maple command `fsolve`. To five decimal places, the solution is

$$m = 0.62006. \tag{11}$$

Of course, using graphics software only tells us that there is one root of the equation on the interval in which we looked. For this problem, proving the uniqueness of a root is done by showing that the function corresponding to the RHS is monotonic increasing — i.e. its derivative is always positive.

Summary

We have seen one technique for solving moving boundary problems involving freezing (or melting): the Boltzmann similarity solution. The Boltzmann similarity solution leads to a **transcendental equation** that needs to be solved graphically (or using numerical methods). However, it only works for problems without a natural length scale. An approximate technique, that of finding the pseudo-steady-state solution is discussed in the next section.

2.4 The pseudo-steady-state approximate solution

A technique is introduced for finding an approximate solution for a moving boundary problem when heat conduction dominates latent heat. (This section may be optionally omitted: the case study does not depend on it.)

The use of the Boltzmann transformation worked well here, and works well for some other problems (see exercises, Question 14, Question 15 and Question 16). It won't work when we introduce a length scale into the problem. Also, it doesn't work in other geometries, such as for cylinders and spheres, where the heat equation takes a different form (so the Boltzmann similarity solution is not valid)†.

One approach for these problems is to find a *pseudo-steady-state solution*, which is an approximate solution obtained by assuming that latent heat release (freezing) is slow compared to conduction of heat. It is obtained by neglecting the time derivative in the heat equation, leaving a simple ordinary differential equation to solve. The following example shows how to do this.

Example 1: *Find the pseudo-steady-state approximation for the heat equation*

$$\frac{\partial u}{\partial t} = \frac{\partial^2 u}{\partial x^2} \tag{1}$$

and the boundary conditions

$$u(0,t) = -1, \qquad u(s(t),t) = 0,$$

with the Stefan condition

$$\frac{\partial u}{\partial x}(s(t),t) = \frac{ds}{dt}.$$

(This is the example problem of Section 2.3.)

Solution: *For the pseudo-steady-state solution we neglect the time derivative in the heat equation, so we solve*

$$\frac{\partial^2 u}{\partial x^2} = 0.$$

Integrating twice gives

$$u(x,t) = c_1(t)x + c_2(t)$$

---

† While the spherical heat equation can be easily changed into the standard heat equation by a simple transformation (see Question 17 of the problems in Chapter 1), this transformation doesn't work with the boundary conditions we would apply on the moving boundary.

where c_1 and c_2 are arbitrary functions of time.

Applying the boundary conditions gives the equations

$$c_2(t) = -1, \qquad c_1(t)s(t) + c_2(t) = 0.$$

Hence,

$$u(x, t) = \frac{x}{s(t)}.$$

Substituting this into the Stefan condition gives

$$\frac{ds}{dt} = \frac{1}{s(t)}$$

which is a separable first-order differential equation. The solution, satisfying $s(0) = 0$, is

$$s(t) = \sqrt{2t}.$$

The pseudo-steady-state solution is applicable when the term $\partial u / \partial t$ in the heat equation is small compared to the $\alpha \partial^2 u / \partial x^2$ term. Physically, this says that latent heat released (for a freezing substance) is mostly conducted away, rather than going to raising the temperature. Scaling arguments can be used (see Question 22) to show this approximation is valid provided the ratio of specific latent heat λ to the heat capacity cu_0 is small, i.e. the dimensionless ratio $\lambda/(\rho c)$ is small.

The pseudo-steady-state approach involves neglecting the time derivative in the heat equation. It is useful for problems where the specific latent heat λ is large compared to the overall heat capacity $c\Delta u$. Some more sophisticated approximate methods, related to the pseudo-steady-state approach are discussed in Hill and Dewynne (1990) and Hill (1987).

2.5 Solving the continuous casting case study

We now use the techniques developed earlier to solve the case-study problem. We use the Boltzmann similarity solution, which leads to the requirement of solving a transcendental equation.

Review of the problem

Our aim is to predict the time that it takes molten steel to solidify a distance $h = 10\,\text{mm}$ when it is poured onto a rotating drum which is rotating so that its surface moves with speed $V = 1\,\text{m/s}$, see Figure 2.1.1. We do not know the temperature of the surface of the copper drum but we know that the temperature at the core is $u_d = 150\,^\circ\text{C}$. The puddle length is given by $V t_h$ where V is the speed of the copper drum surface and t_h is the time it takes the molten steel to solidify the thickness h of the steel sheet.

The temperature in the copper drum u_1 and the temperature of the solidified steel u_2 satisfy

$$\frac{\partial u_1}{\partial t} = \alpha_1 \frac{\partial^2 u_1}{\partial x^2}, \qquad -\infty < x < 0,$$
$$\frac{\partial u_2}{\partial t} = \alpha_2 \frac{\partial^2 u_2}{\partial x^2}, \qquad 0 < x < s(t). \tag{1}$$

The boundary conditions are

$$u(-\infty, t) = u_d,$$
$$u_1(0, t) = u_2(0, t), \tag{2}$$
$$u_2(s(t), t) = u_f.$$

Here u_d is the temperature of the copper a long way from the drum surface, u_f is the solidification temperature of molten steel. We have the continuity of flux condition at $x = 0$,

$$-k_1 \frac{\partial u_1}{\partial x}(0, t) = -k_2 \frac{\partial u_2}{\partial x}(0, t), \tag{3}$$

where k_1 and k_2 are the conductivities of copper and solid steel, respectively. We also have the Stefan condition,

$$k_2 \frac{\partial u_2}{\partial x}(s(t), t) = \rho_2 \lambda \frac{ds}{dt}, \tag{4}$$

where ρ_2 is the density of the solidified steel and λ is the specific latent heat (latent heat per unit mass) released when the molten steel solidifies.

Equivalent boundary conditions

For problems with two regions where the temperature is continuous at the boundary, a useful way to proceed is to decouple the two regions. This is done by introducing an extra unknown, the temperature of the join, U, where

$$u_1(0,t) = u_2(0,t) = U. \tag{5}$$

We can then obtain $u_1(x,t)$ and $u_2(x,t)$ in terms of the unknown U and later use the continuity of flux condition to find the value of U.

It is not immediately obvious that such a solution with constant temperature at the copper-steel interface exists. However, it follows from Sections 2.2 and 2.3 that, with this boundary condition, the heat equation in each of the two materials reduces to a second order ordinary differential equation. We shall see that the four coefficients in the general solution, plus the unknown temperature, U, and the position of the moving boundary, may be uniquely determined from the six boundary conditions.

We now have two sets of equations, defined for each region and linked by the variable U. For the copper drum, $-\infty < x < 0$, we have

$$\frac{\partial u_1}{\partial t} = \alpha_1 \frac{\partial^2 u_1}{\partial x^2}, \qquad u_1(-\infty, t) = u_d, \quad u_1(0,t) = U. \tag{6}$$

For the solidified steel, $0 < x < s(t)$, we have

$$\frac{\partial u_2}{\partial t} = \alpha_2 \frac{\partial^2 u_2}{\partial x^2}, \qquad u_2(0,t) = U, \quad u_2(s(t), t) = u_f. \tag{7}$$

The remaining two boundary conditions we need are the continuity of flux condition (3) which is used to determine U, and the Stefan condition (4) which is used to determine the position of the moving boundary.

Using the Boltzmann similarity solution

The Boltzmann similarity solution for both partial differential equations is

$$u_1(x,t) = C_1 \operatorname{erf}(x/\sqrt{4\alpha_1 t}) + C_2, \tag{8}$$

$$u_2(x,t) = C_3 \operatorname{erf}(x/\sqrt{4\alpha_2 t}) + C_4, \tag{9}$$

where C_1, C_2, C_3 and C_4 are arbitrary constants. These constants will be determined by applying the boundary conditions. Note that we have different values of the thermal diffusivity α_1 and α_2, corresponding to the copper and the steel.

The boundary condition $u_2 = u_f$ on $x = s(t)$ determines the general form of the moving boundary. Applying this condition forces $s(t)$ to be proportional to $t^{1/2}$ so that $\operatorname{erf}(s(t)/\sqrt{4t})$ is a constant. We write

$$s(t) = m\sqrt{4\alpha_2 t}$$

where m is a constant.

We now apply the prescribed temperature boundary conditions at $x = -\infty$, $x = 0$ and $x = s(t)$ given in (6) and (7). Applying these (the calculations are left as an exercise, see Question 19), we obtain

$$u_1(x,t) = U + (U - u_d)\operatorname{erf}(x/\sqrt{4\alpha_1 t}),$$

$$u_2(x,t) = U + (u_f - U)\frac{\operatorname{erf}(x/\sqrt{4\alpha_2 t})}{\operatorname{erf}(m)}.$$

It is easy to quickly check that this satisfies the appropriate boundary conditions. However, these solutions are still in terms of two unknown constants, U and m, which we need to determine.

Applying the remaining boundary conditions

To find U we apply the continuity of flux condition (3) (see exercises, Question 19, for details), and the following equation is obtained for U:

$$U = \frac{u_d \operatorname{erf}(m) + \beta u_f}{\operatorname{erf}(m) + \beta} \quad \text{where} \quad \beta = \frac{k_2}{k_1}\sqrt{\frac{\alpha_1}{\alpha_2}}. \tag{10}$$

Applying the Stefan condition (4) gives a transcendental equation for m (the details are left to the exercises, see question Question 20)

$$me^{m^2}\left(\text{erf}(m) + \beta\right) = \sigma \qquad \text{where} \quad \sigma = \frac{k_2(u_f - u_d)}{\rho_2\lambda\alpha_2\sqrt{\pi}}. \qquad (11)$$

Calculating the puddle length

For this problem we are given the data in Table 2.5.1. Copper is the material of choice for the drum because of its high thermal conductivity k_1, 20 times that of steel. From this data we obtain $\beta = 0.25$ and $\sigma = 1.718$ from equations (10) and (11).

Table 2.5.1. *Data for continuous casting of steel on a copper drum, from Barton (1985).*

h	$10\,\text{mm} = 0.01\,\text{m}$	α_1	$10^{-4}\,\text{m}^2\,\text{s}^{-1}$
V	$1\,\text{ms}^{-1}$	α_2	$4 \times 10^{-6}\,\text{m}^2\,\text{s}^{-1}$
u_f	$1400\,^{\circ}\text{C}$	k_1	$400\,\text{W}\,\text{m}^{-1}\,^{\circ}\text{C}^{-1}$
u_d	$150\,^{\circ}\text{C}$	k_2	$20\,\text{W}\,\text{m}^{-1}\,^{\circ}\text{C}^{-1}$
λ	$2.7 \times 10^5\,\text{J}\,\text{kg}^{-1}$	ρ_2	$7.6 \times 10^3\,\text{kg}\,\text{m}^{-3}$

Solving the transcendental equation (11) numerically we come up with the value of the root $m \simeq 0.839$. Using equation (10) we subsequently obtain the value $U \simeq 460\,^{\circ}\text{C}$ for the temperature of the surface of the drum. This explains why the copper does not melt, even though its melting point is lower than that of steel.

To calculate the time t_h for the molten metal to solidify a distance h we need to determine the time $t = T$ for the moving boundary to be at $x = h$. Using $s(t) = m\sqrt{4\alpha_2 t}$ we obtain

$$t_h = \frac{h^2}{4\alpha_2 m^2}.$$

This is the time for the molten material to solidify a distance h. The puddle length ℓ is

$$\ell = V t_h = \frac{V h^2}{4\alpha_2 m^2}.$$

The above data thus gives an estimate† for the puddle length ℓ as

$$\ell \simeq 9\,\text{metres}.$$

This value is very large compared to the size of the drum, meaning the molten steel will drain off the drum before solidifying. Thus for this method of casting metal to be effective a drum of an impractical size would be needed so that the metal strip can be removed from the drum only after it has completely solidified. This calculation indicates that it is not feasible to cast sheet-steel strips of this thickness by the method of pouring the molten steel onto a cooled rotating drum.

However, the same mathematical model applies to continuous casting of steel slabs between curved channels. Since the case study problem of this Chapter was considered at the 1985 Mathematics-in-Industry workshop, many steel manufacturers, including BHP, have set up continuous slab casters. based on rotating drums. The slabs are thicker (around 10 cm) and the rotation speeds are much slower (around 1 m/min) but the curved channels are indeed of the order of 10 m in length. In fact, a flux powder separates the solidifying steel from the copper channels and this generates additional mathematical modelling tasks.

Further reading

This case study was sourced from Barton (1985). Another quite comprehensive treatment of this problem is given in Fowkes and Mahony (1994). This study also presents some additional calculations, including an analysis of the thickness of the drum and a calculation of the benefit of enhancing cooling by direct water contact. They also briefly discuss an alternative continuous casting arrangement, with two rotating drums (to increase the heat removal and thus speed up the solidification). Shamsi and Mehrotra (1993) use computational techniques to analyse a single drum continuous caster.

A good general reference on moving boundary problems in heat conduction is Hill and Dewynne (1990) which also gives detailed analysis of the pseudo-steady-state approximate method, as discussed in Section 2.4, and introduces some further analytical approximation methods,

† The original article Barton (1985) used a slightly higher value of α_2, $\alpha_2 = 5 \times 10^{-6}\,\text{m}^2\,\text{s}^{-1}$, and obtained an estimate for the puddle length of 7.7 m. However, other calculations developed in this report used the value $\alpha_2 = 5 \times 10^{-6}\,\text{m}^2\,\text{s}^{-1}$, which we have also used here to obtain the puddle length estimate of 9 m.

as does Hill (1987). Tayler (1986) and Fowler (1997) contain advanced treatments of melting and solidification problems which consider "mushy zones" of both liquid and solid phases, simultaneously, for alloy solidification. The authoritative reference for solutions of the heat equation, including problems with moving boundaries, is Carslaw and Jaeger (1959). For diffusion problems, with moving boundaries, see Crank (1975).

Geiger and Poirier (1980) give a metallurgical perspective to melting and solidification problems, including continuous casting. They also treat problems in mass transport, with a moving boundary, including the carbonisation of steel (involving the diffusion of carbon into iron) and the formation of tarnish layers (oxygen diffusing into a metal). Moving boundary problems also occur in the food industry (freezing of food, cooking). See McGowan and McGuinness (1996) for an application to the gelatinisation of cereal starch. Moving boundary problems, of a different form, occur in mathematical finance, such as the valuation of American style options on shares. These are discussed, and compared with the Stefan problem, by Wilmott et al. (1995) and Wilmott (1998).

2.6 Problems for Chapter 2

1. *Suppose the region $x > 0$ is initially liquid at constant temperature u_0. Here u_0 is greater than the freezing temperature u_f. The surface $x = 0$ is then maintained at constant temperature u_1, which is less than the solidification temperature.*

(a) *In which direction does the heat flow?*

(b) *If $u_S(x,t)$ denotes the temperature in the solid phase, and $u_L(x,t)$ denotes the temperature in the liquid phase, then deduce the Stefan condition*

$$-k_L \frac{\partial u_L}{\partial x}(s(t),t) + k_S \frac{\partial u_S}{\partial x}(s(t),t) = \rho\lambda\frac{ds}{dt}$$

where k_S and k_L are the respective conductivities.

(c) *What are the other boundary conditions for this problem?*

2. *A solid is initially everywhere at the melting temperature u_m. One end, $x = 0$, is such that the temperature at that end is always at a constant temperature $u_1 > u_m$. The solid then melts from left to right.*

(a) *In which direction does the heat flow?*

(b) Derive the Stefan condition for this problem,

$$-k\frac{\partial u}{\partial x}(s(t),t) = \rho\lambda\frac{ds}{dt},$$

where $u(x,t)$ is the temperature of the liquid phase.

3.   A **super-cooled** liquid occurs when the liquid phase exists at below the freezing point u_f.

Suppose that the region $x > 0$ initially contains liquid at temperature $u_1 < u_f$ and that solidification starts at $x = 0$ and moves to the right, no heat being removed from the solidified material whose temperature will thus have constant value u_f throughout.

Derive the Stefan condition,

$$-k\frac{\partial u}{\partial x}(s(t),t) = \rho\lambda\frac{ds}{dt}.$$

4.   Consider the problem of the freezing of a sphere (e.g. in the food industry, the freezing of the water content of a pea or orange). The govering partial differential equation is the radially symmetric heat equation in spherical coordinates

$$\frac{\partial u}{\partial t} = \frac{\alpha}{r^2}\frac{\partial}{\partial r}\left(r^2\frac{\partial u}{\partial r}\right)$$

where $u(r,t)$ is the temperature at time t and at a distance r from the centre of the sphere and α is the thermal diffusivity.

If the solidification front is at $r = s(t)$, determine the Stefan condition for this problem. [Hint: Consider the case of the sphere freezing from $r = s(t)$ to $r = s(t + \delta t)$.]

5.   Using the Boltzmann similarity solution, solve the following problem:

$$\frac{\partial u}{\partial t} = 4\frac{\partial^2 u}{\partial x^2} \qquad x \geq 0$$

with the boundary conditions

$$u(0,t) = 0, \qquad u(\infty,t) = 10.$$

6.   A field of crops has been sprayed with a pesticide. The wind picks up some of the pesticide and carries it along. There is also diffusion of the pesticide caused by turbulent fluctuations, mainly in the y direction. Assume

$$V\frac{\partial c}{\partial x} = D\frac{\partial^2 c}{\partial y^2}$$

where $c(x,y)$ is the steady state concentration at a height y and a distance x from the start of the field (see Figure 2.6.1). Also, V is the (constant) velocity of the wind and D is the diffusivity. Assume the domain of the problem is a quarter plane ($x > 0$, $y > 0$).

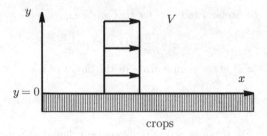

Fig. 2.6.1. Diagram for Question 6.

(a) Given that the concentration of the pesticide at ground level remains constant, solve for $c(x, y)$ in terms of error functions. [Hint: Use the Boltzmann similarity solution written in terms of the variables in this problem.]

(b) Give a sketch of the concentration at different times.

7. Show that the substitution

$$u(x, t) = f(\eta), \qquad \eta = xt^{-1/2}$$

reduces the partial differential equation

$$\frac{\partial u}{\partial t} = \alpha \frac{\partial^2 u}{\partial x^2}$$

to an ordinary differential equation. (Just obtain the differential equation, don't solve it.)

8. Show that the substitution

$$u(x, t) = f(\eta), \qquad \eta = xt^{-1/2}$$

does not reduce the wave equation

$$\frac{\partial^2 u}{\partial t^2} = \frac{\partial^2 u}{\partial x^2}$$

to an ordinary differential equation. What about the substitution

$$u(x, t) = f(\eta), \qquad \eta = xt^{-1}?$$

9. We showed that the Boltzmann similarity transformation $u(x, t) = f(\eta)$, $\eta = x/\sqrt{\alpha t}$ reduces the heat equation to an ordinary differential equation. Suppose we have the heat equation on a finite domain, $0 < x < 1$, subject to the boundary conditions

$$u(0, t) = 100, \quad u(1, t) = 200.$$

Explain why the Boltzmann similarity transformation will not work on this problem.

10.   Show that the **only** substitution of the form $\eta = xt^\beta$ which reduces the heat conduction equation

$$\frac{\partial u}{\partial t} = \alpha \frac{\partial^2 u}{\partial x^2}$$

to an ordinary differential equation is when $\beta = -1/2$. (i.e. substitute $u(x,t) = f(xt^\beta)$ into the heat equation and hence deduce that $\beta = -1/2$ is the only possibility for eliminating t and x.)

11.   Tarnishing. The metal zirconium absorbs a significant amount of oxygen during the process of oxidation. Once the oxide layer has formed its thickness $h(t)$ increases according to

$$h(t) = \frac{M_1}{M_2} \sqrt{4k't}$$

where k' is a positive constant, t is time and M_1 and M_2 are the molar volumes of Zr and $Zr\,O_2$ respectively; Geiger and Poirier (1980).

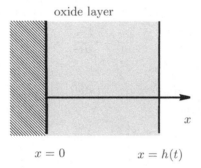

oxide layer

$x = 0$ $x = h(t)$

Fig. 2.6.2. Tarnishing of metal by an oxidised layer. See Question 11.

The oxygen ions diffuse from the air into the oxidised layer according to the diffusion equation

$$\frac{\partial C}{\partial t} = D \frac{\partial^2 C}{\partial x^2}$$

where $C(x,t)$ is the concentration of oxygen ions (mols/cm^3).

(a) Given the concentration at the interface $x = h(t)$ is the equilibrium concentration C_e, and the initial concentration is zero, with $C(\infty, t) = 0$, use the Boltzmann similarity solution to deduce

$$C(x,t) = C_e \frac{1 - \operatorname{erf}(x/\sqrt{4Dt})}{1 - \operatorname{erf}\left(\dfrac{M_1}{M_2}\sqrt{\dfrac{k'}{D}}\right)}.$$

(b) Where is the oxygen concentration a maximum?

12. A transcendental equation is an equation which does not have an algebraic solution. One example is the equation

$$\sin(m) = \frac{1}{2}m.$$

Discuss how many solutions exist and find these solutions graphically.

13. Prove that the transcendental equation

$$m\,\mathrm{erf}(m)e^{m^2} = \sigma,$$

where σ is a positive constant, always has a unique solution, for $\sigma > 0$. [Hint: think graphically! Show that the LHS is a monotonic increasing function of m. What else is needed?]

14. The region $x > 0$ is initially liquid just at the solidification temperature u_f. The end $x = 0$ is maintained at a temperature u_1 where $u_1 < u_f$. The liquid then freezes from $x = 0$ to the right.

The temperature $u(x, t)$ of the solid part satisfies the heat equation

$$\frac{\partial u}{\partial t} = \alpha\frac{\partial^2 u}{\partial x^2}$$

with the boundary conditions

$$u(0, t) = u_1, \qquad u(s(t), t) = u_f$$

and the Stefan condition

$$k\frac{\partial u}{\partial x} = \rho\lambda\frac{ds}{dt}.$$

(a) Explain why we assume $s(t) = m\sqrt{4\alpha t}$.

(b) Hence find the temperature and also show that m satisfies the equation

$$m\,\mathrm{erf}(m)e^{m^2} = \sigma, \qquad \sigma = \frac{k(u_f - u_1)}{\rho\lambda\alpha\sqrt{\pi}}.$$

(c) Hence determine how long it takes to freeze water a distance of $10\,\mathrm{cm}$, if $u_1 = -10\,^\circ\mathrm{C}$.

15. Suppose the region $x > 0$ is initially liquid at constant temperature u_0. The surface $x = 0$ is then maintained at constant temperature u_1 (which is less than the melting temperature u_m).

The temperature of the solid and liquid phases, $u_S(x, t)$ and $u_L(x, t)$, satisfy

$$\frac{\partial u_S}{\partial t} = \alpha_S\frac{\partial^2 u_S}{\partial x^2}, \qquad \frac{\partial u_L}{\partial t} = \alpha_L\frac{\partial^2 u_L}{\partial x^2},$$

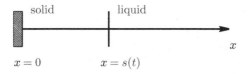

Fig. 2.6.3. Diagram for Question 15.

with $u_S(0, t) = u_1$ and $u_L(x, 0) = u_0$. On the moving boundary $u_L = u_S = u_f$, where u_f is the freezing temperature. We also have the Stefan condition

$$-k_L \frac{\partial u_L}{\partial x}(s(t), t) + k_S \frac{\partial u_S}{\partial x}(s(t), t) = \rho\lambda\frac{ds}{dt}.$$

(a) Show that the solution may be written as

$$u_S(x, t) = u_1 + \frac{u_f - u_1}{\operatorname{erf}(m)}\operatorname{erf}(x/\sqrt{4\alpha_S t}),$$

$$u_L(x, t) = u_0 - \frac{(u_0 - u_f)}{1 - \operatorname{erf}(m\sqrt{\alpha_S/\alpha_L})}\left(1 - \operatorname{erf}(x/\sqrt{4\alpha_L t})\right).$$

(b) Give the transcendental equation satisfied by m.

16. A **super-cooled** liquid occurs when the liquid phase exists at below the freezing point u_f.

Suppose that the region $x > 0$ initially contains liquid at temperature $u_1 < u_f$ and that solidification starts at $x = 0$ and moves to the right, no heat being removed from the solidified material the temperature of which will thus have constant value u_f throughout.

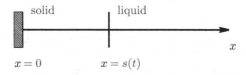

Fig. 2.6.4. Diagram for Question 16.

(a) Find the temperature distribution in the liquid phase and show that it depends on the solution of the transcendental equation

$$me^{m^2}(1 - \operatorname{erf}(m)) = \sigma$$

where m is a parameter related to the position of the moving boundary and σ is a constant which depends on the thermal properties of the material.

(b) Show, that for large values of m the left hand side of the transcendental equation approaches an asymptote. [Hint: use a mathematical handbook (e.g. Spiegel (1968)) to find a suitable expansion for erf for large values of its argument.] Hence discuss the existence of a solution of the super-cooling problem for different values of the right hand side of the transcendental equation.

17.   Consider the problem of a solidifying semi-infinite liquid which is initially at the melting temperature u_m. Suppose that instead of the boundary $x = 0$ having a prescribed temperature, instead it satisfies a Newton cooling condition.

(a) Write down all the appropriate governing equations.

(b) Show that the Boltzmann transformation will not work here.

(c) Find a pseudo-steady-state approximate solution (see Section 2.4), by neglecting the time derivative in the heat equation.

18.   A cylindrical pipe, of radius $b = 5\,\mathrm{cm}$, contains water which is initially at the freezing temperature $u_f = 0\,°\mathrm{C}$. The metal surface of the cylinder is held at a temperature $u_1 = -10\,°\mathrm{C}$, which is below the freezing temperature. A moving front starts from the metal surface and moves inward.

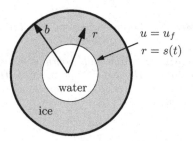

Fig. 2.6.5. Freezing inside a cylindrical pipe. Diagram for Question 18.

(a) Find a pseudo-steady-state solution (by neglecting the time derivative in the heat equation — see Section 2.4).

(b) Hence calculate the time taken for all the water inside the cylinder to freeze.

19.   For the continuous casting problem in Section 2.5 the temperature in the drum u_1 and the temperature in the solid part of the metal being cast u_2 are given by

$$u_1(x,t) = U + (U - u_d)\,\mathrm{erf}(x/\sqrt{4\alpha_1 t}), \qquad x \le 0,$$

$$u_2(x,t) = U + (u_f - U)\frac{\mathrm{erf}(x/\sqrt{4\alpha_2 t})}{\mathrm{erf}(m)}, \qquad 0 \le x \le m\sqrt{4\alpha_2 t}.$$

Apply the continuity of flux condition to determine the constant U.

20.   (Continuation of previous question.) Apply the Stefan condition to the temperatures to obtain a transcendental equation for the constant m. Show that this equation may be written in the form

$$me^{m^2}\left(\beta + \operatorname{erf}(m)\right) = \frac{k_2(u_f - u_d)}{\rho_2 \lambda \alpha_2 \sqrt{\pi}}$$

where $\beta = k_2\sqrt{\alpha_1}/(k_1\sqrt{\alpha_2})$.

21.   Under high temperatures steel can undergo a process of decarburisation where the carbon in the steel diffuses out of the steel (see Geiger and Poirier (1980)). It does this in such a way that an interface is formed where the carbon concentration is discontinuous across the interface. Over time the interface moves further into the material, causing a layer of weaker strength steel. A typical concentration profile is shown in Figure 2.6.6.

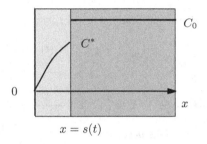

Fig. 2.6.6. Carbon concentration profile in steel during the process of decarburising. See Question 21.

The problem is to calculate the depth of decarburisation of initially $C_0 = 0.2\%$ carbon steel after 1 year of exposure to severe decarburising conditions at $900\,^{\circ}\mathrm{C}$. At this temperature the diffusivity of carbon in steel is $10^{-9}\ \mathrm{cm}^2/\mathrm{sec}$. In the decarburised region the carbon concentration at the interface is always $C^* = 0.02\%$ whereas the carbon concentration at the surface is zero.

This problem has similarities to the solidification problems discussed in this chapter although we are considering mass transport (of carbon atoms) rather than heat transport. In the decarburised region, the carbon concentration $C(x,t)$ satisfies the diffusion equation

$$\frac{\partial C}{\partial t} = D\frac{\partial^2 C}{\partial x^2}.$$

(a) Consider a small time interval where the interface advances and apply a **mass** balance over that time interval to deduce that

$$(C_0 - C^*)\frac{ds}{dt} = D\frac{\partial C}{\partial x}(s(t), t), \tag{1}$$

a condition at the moving boundary analogous to the Stefan condition in heat transport.

(b) Hence calculate the concentration profile for $x < s(t)$, assuming $s(t) = m\sqrt{4Dt}$. (Why can you assume this?)

(c) Apply the condition (1) to obtain a transcendental equation for m,

$$m \operatorname{erf}(m) e^{m^2} = \frac{1}{\sqrt{\pi}} \left(\frac{C^*}{C_0 - C^*} \right).$$

(d) Using the data given above, obtain the thickness of the decarburised layer after one year.

22.   For the simplest Stefan problem, as considered in Section 2.3, the temperature $u(x, t)$ satisfies the heat equation

$$\frac{\partial u}{\partial t} = \alpha \frac{\partial^2 u}{\partial x^2},$$

and the moving boundary condition

$$\frac{ds}{dt} = k \frac{\partial u}{\partial x} \qquad \text{on } x = s(t)$$

(a) Using the scaling

$$U = \frac{u}{u_0}, \quad X = \frac{x}{\ell}, \quad T = \frac{t}{\ell^2/\alpha}, \quad S = \frac{s}{\ell}$$

show that the equations become

$$\frac{\partial U}{\partial T} = \frac{\partial^2 U}{\partial X^2}, \qquad \text{with} \quad \frac{dS}{dT} = \epsilon \frac{\partial U}{\partial X} \quad \text{on } X = S(T)$$

and define the dimensionless constant ϵ.

(b) For small ϵ the boundary moves very slowly. We can define a new time τ by $T = \epsilon \tau$. Write the equations in terms of τ, and then let $\epsilon \to 0$. Show that the resulting equations correspond to the pseudo-steady-state approximation.

(c) In a couple of lines, explain what physical quantity in the governing equations is being neglected by the approximation $\epsilon \to 0$.

3

Case Study: Water Filtration

We consider a problem involving filtration of water where the aim is to determine the salt concentration build up in a water filter after it has been operating for some time. To solve the PDE for the model we develop we extend the technique of Chapter 2 by introducing the method of stretching transformations, which is a technique for obtaining similarity solutions of PDEs. The basic idea is to look for stretching symmetries of the PDEs and boundary conditions which allow the construction of variable combinations which reduce the PDE into an ODE.

3.1 Introduction to the case study problem

In this section we explore a simple model which describes the removal of salt from solution through a semi-permeable membrane. The process is called reverse osmosis. This case study is based on material in Probstein (1989).

Problem background

It is vitally important to many industries to be able to effectively filter impurities from liquids. One example of a filtration process is the purification of water by removal of salt (see, for example, James et al. (1993)).

One method for effecting filtration is to pass the water along a semi-permeable membrane which allows the passage of water but not the salt ions. In practice a filtering mechanism is constructed by stacking parallel

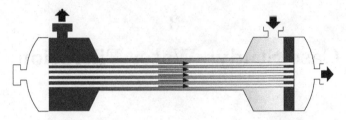

Fig. 3.1.1. A filter designed to extract pure water from contaminated water or salt water. Salt water passes parallel to a semi-permeable membrane with pure water passing through the membrane.

flat semi-permeable membranes separated by narrow gaps or by using bundles of hollow fibre circular tubes. The pressure forces pure water through the semi-permeable membrane.

Without pressure driving the flow there would be a tendency for water to flow in the opposite direction to cause the concentrated mixture to become diluted. This process is called ***osmosis***. The application of pressure drives pure water through the semi-permeable membrane; this process is called reverse osmosis. Pressure also forces the salt-water mixture along the duct. The water flux through the membrane is proportional to the difference between the pressure in the solution and the osmotic pressure difference.

The osmotic pressure difference is proportional to the salt concentration, so that any build-up of salt near the semi-permeable membrane will reduce the efficiency of the filtering system. This means the filter will have to be shut down occasionally and back flushed to remove the salt build-up on the membrane before the filter can operate again.

It is therefore desirable to create a mathematical model which can predict the salt concentration on the membrane as a function of the distance along the membrane. This gives the designers information on how long the filter can be.

Model problem

We are interested in the salt concentration ***very close*** to the semi-permeable membrane, since this is where salt build-up occurs. A simple model of this is shown below in Figure 3.1.2. We consider a semi-infinite region $x > 0$, $y > 0$ with the semi-permeable membrane at $y = 0$. Fluid flows through the duct with speed $v(y)$, parallel to the x-axis. At the membrane, pressure forces the water out at a known flow rate q.

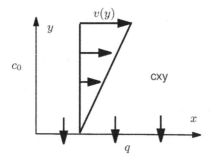

Fig. 3.1.2. Diffusion and advection near a semi-permeable membrane.

The mathematical model must account for diffusion and advection. If we define a concentration $C(x, y, t)$ then the advection-diffusion equation is

$$\frac{\partial C}{\partial t} + v_1(y)\frac{\partial C}{\partial x} = D\left(\frac{\partial^2 C}{\partial x^2} + \frac{\partial^2 C}{\partial y^2}\right). \tag{1}$$

Here $v_1(y)$ is the horizontal component of velocity, which is dependent on distance from the membrane — it is zero on the membrane surface.

Since the salt concentration varies rapidly in the y-direction we would expect the diffusion of salt to be mainly in the y-direction and thus neglect diffusion in the x-direction. If we also restrict attention to equilibrium concentrations, setting $\partial C/\partial t = 0$, the governing equation (1) reduces to

$$v_1(y)\frac{\partial C}{\partial x} = D\frac{\partial^2 C}{\partial y^2}. \tag{2}$$

On the membrane, the tangential velocity of the saline mixture will be zero. Above the membrane the velocity will increase to some typical velocity v_0 through the duct. We model the velocity in the duct, near the semi-permeable membrane, by

$$v(y) = \frac{v_0 y}{h} \tag{3}$$

where h is the distance from the semi-permeable boundary to the centre of the duct. Note that this horizontal velocity is zero on the semi-permeable membrane. The variation of velocity with distance from the

membrane is important for suitable vertical concentration gradients to
be set up. The governing differential equation for the salt concentration
is thus

$$y\frac{\partial C}{\partial x} = \alpha\frac{\partial^2 C}{\partial y^2}, \qquad \text{where} \quad \alpha = \frac{Dh}{v_0}. \qquad (4)$$

Since we are interested in the salt concentration very close to the
semi-permeable membrane we can consider a semi-infinite region $y > 0$
where the concentration far from the membrane is the concentration of
the solution before it flows into the filter, c_0. Thus

$$C(0, y) = c_0 \quad \text{and} \quad C(x, \infty) = c_0. \qquad (5)$$

We also need a boundary condition on $y = 0$.

Since no salt flows through the semi-permeable membrane then there
is no mass flux of salt at $y = 0$. However, since there is a vertical velocity
at $y = 0$ then we have two components to the salt mass flux — the flux
due to diffusion J_{diff} and the flux due to advection J_{adv},

$$J = J_{\text{diff}} + J_{\text{adv}}. \qquad (6)$$

The flux due to diffusion is given by Fick's law

$$J_{\text{diff}} = -D\frac{\partial C}{\partial y}. \qquad (7)$$

The flux due to advection is

$$J_{\text{adv}} = -qC(x, 0) \qquad (8)$$

where q is the flow rate of water through the membrane. There is a
minus sign because the water flow is in the opposite direction to the
y-axis. Since $J(x, 0) = 0$ on the membrane surface, $y = 0$, then the
boundary condition on the membrane surface is

$$-D\frac{\partial C}{\partial y}(x, 0) - qC(x, 0) = 0. \qquad (9)$$

Excess concentration

We anticipate the overall concentration will not vary much from the inlet
concentration c_0. Let us therefore define the **excess concentration**
$c(x, y)$ by

$$c(x, y) = C(x, y) - c_0.$$

Substituting this into the governing equations, we obtain

$$y\frac{\partial c}{\partial x} = \alpha\frac{\partial^2 c}{\partial y^2}, \qquad \alpha = \frac{Dh}{v_0} \qquad (10)$$

with the homogeneous boundary conditions,

$$c(0, y) = 0, \quad c(x, \infty) = 0. \qquad (11)$$

We also have, from (9), the boundary condition on the membrane surface,

$$-D\frac{\partial c}{\partial y}(x, 0) = q(c(x, 0) + c_0). \qquad (12)$$

We can further simplify the governing equations by assuming the concentration does not vary by much from the concentration c_0. We thus assume $c(x, 0) \ll c_0$ and the boundary condition (12) may therefore be approximated by

$$-D\frac{\partial c}{\partial y}(x, 0) = qc_0. \qquad (13)$$

Discussion

To solve this problem we aim to reduce the partial differential equation to an ordinary differential equation. We need some transformation, like the Boltzmann transformation used in the previous chapter. For this, we will introduce the method of stretching transformations. The basic idea is to look for a family of simple changes of variable which leave the governing equations invariant (i.e. in exactly the same form in the new variables as in the old). Since the solution of the governing equations must share this property, we can construct similarity variables which transform the PDE into a single ODE in a similarity variable.

In the following two sections, we look at some other problems which will allow us to develop the appropriate technique. The first problem revisits the Boltzmann transformation, using the generalised technique to obtain it. The second shows how to use the technique to come up with a new similarity transformation for the heat equation but with boundary conditions that are incompatible with the usual Boltzmann transformation.

3.2 Stretching transformations

Here we introduce a more general procedure, known as the method of stretching transformations, which gives the appropriate functional form to reduce the given partial differential equation to an ordinary differential equation. The basic idea of the method is to apply a family of stretching transformations of the independent and dependent variables and determine which transformations leave the equations unchanged (or invariant).

Introduction

In Section 2.2 we saw how the Boltzmann transformation expressed the temperature as a function of a single composite variable $x/\sqrt{\alpha t}$. This transformation reduced the heat conduction equation to an ordinary differential equation. What we would like to be able to do is to generalise the Boltzmann similarity solution for the heat equation so that it can be used to solve different partial differential equations.

The way we go about generalising the Boltzmann similarity solution is to consider the effect of transforming the variables on the governing equations. We define a general family of stretching transformations and determine particular stretchings for which the governing equations are the same in the new variables as in the old variables. The fact that the solution must have this same property allows us to construct similarity variables, such as the Boltzmann similarity variable, which reduce a PDE to an ODE.

The general stretching transformation

Let us consider a family of transformations of the variables x, t, C to some new variables x_*, t_*, C_* given by

$$x = e^a x_*, \quad t = e^b t_*, \quad C = e^\gamma C_*, \tag{1}$$

where a, b and γ are three parameters, to be determined. Transformations of the form (1) are called **stretching transformations** since the transformation corresponds to a rescaling, or magnification, of each variable, where e^a, e^b and e^γ are the magnification factors.

The set of transformations (1) forms a **3-parameter** family of transformations. Our aim is to find a subset of transformations which have the property of not changing the governing equations. This is called **invariance**.

In the following we go through three steps:

- Step 1: Substitute the 3-parameter family into the PDE and obtain a condition on the constants.
- Step 2: Obtain the 1-parameter family of transformations for which the governing equations are invariant.
- Step 3: Obtain combinations of variables which are invariant under the 1-parameter family.

Example problem

The governing equations for an example diffusion problem are

$$\frac{\partial C}{\partial t} = \frac{\partial^2 C}{\partial x^2} \tag{2}$$

with boundary conditions

$$C(0, t) = 1, \quad C(\infty, t) = 0 \tag{3}$$

and initial condition

$$C(x, 0) = 0. \tag{4}$$

Step 1: invariance of governing equations

We look for the subset of the 3-parameter set of transformations which leave the governing equations invariant. This is illustrated by the following two examples, the first for the PDE and the second for the boundary conditions.

Example 1: *Find conditions on a, b and c for the PDE (2) to be invariant under the general stretching transformation (1).*

Solution: *We now substitute the general stretching transformation (1) into the partial differential equation (2). Now, since e^a, e^b and e^γ are constants, then†*

$$\frac{\partial C}{\partial t} = \frac{e^\gamma}{e^b}\frac{\partial C_*}{\partial t_*}, \qquad \frac{\partial^2 C}{\partial x^2} = \frac{e^\gamma}{e^{2a}}\frac{\partial^2 C_*}{\partial x_*^2}.$$

Substituting into the PDE, equation (2), we obtain

$$e^{\gamma-b}\frac{\partial C_*}{\partial t_*} = e^{\gamma-2a}\frac{\partial^2 C_*}{\partial x_*^2}. \tag{5}$$

Dividing throughout by the factor $e^{\gamma-2a}$ we obtain

$$e^{2a-b}\frac{\partial C_*}{\partial t_*} = \frac{\partial^2 C_*}{\partial x_*^2}. \tag{6}$$

Whenever $2a-b=0$ we see that (6) is of the same form as (2). We then say that the partial differential equation (2) is invariant under the the stretching transformation (1) if

$$2a - b = 0. \tag{7}$$

This is the most general condition for invariance of the PDE. Now we examine invariance of the boundary and initial conditions. This further narrows the possible choice of invariant combinations of variables.

Example 2: *Determine conditions on the constants a, b and c so that the boundary conditions (3) are invariant under the general stretching transformation (1).*

† Formally, we use the chain rule, $\dfrac{\partial C}{\partial t} = \dfrac{\partial(e^\gamma C_*)}{\partial t^*}\dfrac{\partial t_*}{\partial t}.$

Solution: *The boundary condition (3) can be written as*

$$C = 1 \qquad \text{when } x = 0.$$

Substituting the general 3-parameter stretching transformation (1) this becomes

$$e^\gamma C_* = 1 \qquad \text{when } e^a x_* = 0.$$

While e^a cancels, this boundary condition will only be invariant if

$$\gamma = 0. \tag{8}$$

For the initial condition (4),

$$C = 0 \qquad \text{when } t = 0.$$

We substitute the general stretching transformation (1) obtaining

$$e^\gamma C_* = 0 \quad \text{when } e^a t_* = 0$$

which simplifies to

$$C_* = 0 \quad \text{when } t_* = 0 \qquad \Rightarrow \qquad C_*(x_*, 0) = 0.$$

Thus this boundary condition (4) is always invariant under the general 3-parameter stretching transformation (1). It gives no extra information about the constants a, b or γ.

Finally, the condition at $x = \infty$ is also invariant under the general family of transformations. So this gives no further information about the constants a, b and γ.

Step 2: obtain 1-parameter family of transformations

The overall problem, defined by (2) with (3) and (4), is invariant under the general stretching transformation (1) if

$$2a - b = 0 \quad \text{and} \quad \gamma = 0.$$

This gives two simultaneous equations in three unknowns. We can solve in terms of one† of the variables, a, so that

$$b = 2a \quad \text{and} \quad \gamma = 0. \tag{9}$$

† As a rule of thumb, it is usually best to solve in terms of the constant corresponding to the highest derivative. For this example, it is the constant a, corresponding to the variable x.

Substituting this back into the original family of transformations gives the 1-parameter family of transformations

$$x = e^a x_*, \quad t = e^{2a} t_*, \quad C = C_*. \tag{10}$$

All the governing equations are now invariant under this 1-parameter family of transformations.

Step 3: Obtaining invariant combinations

The overall objective here is to obtain two invariant combinations of the variables, one of which involves the dependent variable u, allowing us to relate the solution to another invariant combination.

Example 3: *Construct combinations of the variables x, t and u, which are invariant under the 1-parameter family of transformations (10).*

Solution: *Eliminating the parameter a, by solving for e^a we obtain*

$$e^a = \frac{x}{x_*} = \left(\frac{t}{t_*}\right)^{1/2}. \tag{11}$$

We can use this to find invariant combinations of the variables.

By cross-multiplying, we obtain an invariant combination of variables

$$x t^{-1/2} = x_* t_*^{-1/2}. \tag{12}$$

The dependent variable u, here, is itself an invariant, since it is unchanged by the transformation (10). This will not always be the case, however, as we see in the next section. We introduce η_1 and η_2 to denote the two invariant combinations

$$\eta_1 = x t^{-1/2} = x_* t_*^{-1/2}, \qquad \eta_2 = C = C_*. \tag{13}$$

Since the partial differential equation and its boundary conditions are invariant under the transformation (10) the unique solution must also have this property. In other words, we must be able to write $\eta_2 = f(\eta_1)$, where f is some function to be determined. This gives

$$C(x,t) = f(x t^{-1/2}). \tag{14}$$

A common error is to forget to divide through by the coefficient of one of the sides of the equation. In equation (5) if we forgot to divide through by $e^{\gamma - 2a}$ then invariance of the partial differential equation could be obtained by setting both $\gamma - b = 0$ and $\gamma - 2a = 0$. However, this does not lead to a sufficiently general set of transformations to construct the invariant combinations.

Reducing the PDE to an ODE

This change of variables (14) can now be substituted back into the PDE. For this we use the chain rule. These details were carried out previously in Section 2.2 (but set $\alpha = 1$). The PDE (2) reduces to the ODE

$$-\frac{1}{2}\eta\frac{df}{d\eta} = \frac{d^2 f}{d\eta^2}. \tag{15}$$

We also saw, in Section 2.2, that this ODE had the general solution

$$f(\eta) = C_1 \operatorname{erf}\left(\frac{\eta}{2}\right) + C_2$$

where C_1 and C_2 are arbitrary constants. These constants can be evaluated by applying the boundary conditions.

The reader may have noticed that the procedure adopted above can lead to other similarity transformations. For example, instead of the invariant combinations (13), we could have come up with

$$\eta_1 = t/x^2, \qquad \eta_2 = C.$$

This leads to the similarity transformation $C = g(t/x^2)$ and the transformation $C = f(\phi)$, $\phi = t/x^2$. This transformation reduces the partial differential (2) to the ordinary differential equation

$$4\phi\frac{d}{d\phi}\left(\phi\frac{dg}{d\phi}\right) + 2\phi\frac{dg}{d\phi} = \frac{dg}{d\phi}.$$

However, this differential equation can be transformed into the differential equation (15) by the change of variables $\phi = 1/\sqrt{\eta}$.

Further generalisations

Although, in these notes, we will use only stretching transformations, it is interesting to know how the approach can be generalised. One may investigate invariance under any 1-parameter transformation group, for example, rotations and translations,

$$\begin{bmatrix} x \\ t \end{bmatrix} = \begin{bmatrix} \cos(a) & -\sin(a) \\ \sin(a) & \cos(a) \end{bmatrix} \begin{bmatrix} x_* \\ t_* \end{bmatrix}, \qquad \begin{bmatrix} x \\ t \end{bmatrix} = \begin{bmatrix} x_* \\ t_* \end{bmatrix} + \begin{bmatrix} a \\ b \end{bmatrix}.$$

The idea is to see if the partial differential equation (and initial and boundary conditions) is invariant under the transformation and if so, to then form invariant combinations of the variables by eliminating the parameters a and b.

Generalising further, a 1-parameter family of transformations is

$$\begin{aligned} x &= G_1(x_*, y_*, C_*; a), \\ t &= G_2(x_*, y_*, C_*; a), \\ C &= G_3(x_*, y_*, C_*; a), \end{aligned} \tag{16}$$

for arbitrary functions G_1, G_2, G_3. It is especially convenient if the parameter value $a = 0$ corresponds to the identity transformation. The basic idea is then to look at an infinitesimal transformation which can be obtained by expanding (16) in a Taylor series and neglecting $O(a^2)$ terms. Invariance of the partial differential equation (and boundary and initial conditions) is then sought under this infinitesimal transformation. This approach is known as the infinitesimal method of Lie groups. This technique, and further generalisations, are useful for finding exact solutions of complicated nonlinear partial differential equations. Elementary accounts of Lie symmetry group methods are found in Dresner (1983) and Hill (1992).

3.3 Diffusion from a point source

We use the stretching transformation method to obtain another physically important solution of the diffusion equation. This solution corresponds to supplying a mass M_0 of a substance to a single point, in an instant of time. This type of solution is called a point source solution.

The problem

Suppose that a mass M_0 of some pollutant is suddenly released into the sea, concentrated at a point. Our aim is to determine the subsequent concentration of the pollutant at various distances from where it was released. By symmetry, the pollutant diffuses in the radial direction, as in Figure 3.3.1.

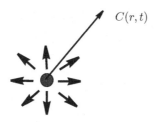

$C(r,t)$

Fig. 3.3.1. A mass source applied to $r = 0$ at time $t = 0$.

Governing equations

The governing partial differential equation is the diffusion equation in spherical coordinates

$$\frac{\partial C}{\partial t} = D \left(\frac{1}{r^2} \frac{\partial}{\partial r} \left(r^2 \frac{\partial C}{\partial r} \right) \right)$$

where $C(r,t)$ is the pollutant concentration and D is the diffusivity of the pollutant in water. The right-hand side of this equation is $\nabla^2 C$ expressed in spherical coordinates and neglecting any dependence on any variable other than radial distance r.

We would expect that the concentration a long distance from where the pollution was released would be zero, hence one boundary condition is

$$C(\infty, t) \to 0.$$

To obtain an initial condition we make use of the information that a mass M_0 of the pollutant is released initially. Since the concentration is the ratio of the mass of the pollutant to the overall volume of the mixture, the total mass of pollutant, for time $t > 0$, is given by

$$M_0 = 4\pi \int_0^\infty C(r,0)\, r^2\, dr.$$

For mathematical simplicity we shall set $D = 1$, $M_0 = 1$. (This is equivalent to scaling the equations.) The governing equations are now

$$\frac{\partial C}{\partial t} = \frac{1}{r^2}\frac{\partial}{\partial r}\left(r^2\frac{\partial C}{\partial r}\right) \tag{1}$$

with the boundary condition

$$C(\infty, t) = 0 \tag{2}$$

and the initial condition

$$\int_0^\infty C(r,0)\, r^2\, dr = \frac{1}{4\pi}. \tag{3}$$

We now seek a suitable similarity transformation which reduces the partial differential equation to an ordinary differential equation, which we can hopefully solve.

Stretching transformation

Let us consider the general, 3-parameter, stretching transformation

$$r = e^a r_*, \quad t = e^b t_*, \quad C = e^\gamma C_*, \tag{4}$$

where a, b and γ are constants to be determined. We substitute (4) into the partial differential equation (1) and boundary initial conditions (2) and (3) to determine a particular stretching transformation for which (1), (2) and (3) are invariant. The following example shows the effect of requiring the PDE and the initial condition to be invariant.

Example 1: *Find a similarity variable that reduces the governing equations to an ODE.*

Solution: Substituting (4) into the partial differential equation (1) we obtain

$$e^{\gamma - b} \frac{\partial C_*}{\partial t_*} = e^{\gamma - 2a} \frac{1}{r_*^2} \frac{\partial}{\partial r_*} \left(r_*^2 \frac{\partial C_*}{\partial r_*} \right).$$

Dividing through by $e^{\gamma - 2a}$ gives

$$e^{2a - b} \frac{\partial C_*}{\partial t_*} = \frac{1}{r_*^2} \frac{\partial}{\partial r_*} \left(r_*^2 \frac{\partial C_*}{\partial r_*} \right).$$

Hence the partial differential equation is invariant if and only if

$$2a - b = 0. \tag{5}$$

Finally we substitute the stretching transformation (4) into the initial condition (3). In the integral, we use a change of variable, $r = e^a r_*$ so that $dr = e^a dr_*$, and thus (3) becomes

$$e^{\gamma + 3a} \int_0^\infty C_* \, r_*^2 \, dr_* = \frac{1}{4\pi} \quad \text{when } t_* = 0.$$

This condition is invariant if

$$\gamma + 3a = 0. \tag{6}$$

We now solve for the constants in terms of a (which corresponds to the variable r). For the entire problem to be invariant, $\gamma = -3a$ and $b = 2a$. Thus the general 3-parameter stretching transformation (4) reduces to the 1-parameter stretching transformation

$$r = e^a r_*, \quad t = e^{2a} t_*, \quad C = e^{-3a} C_*. \tag{7}$$

By eliminating a we find invariant combinations of the variables, i.e.

$$e^a = \frac{r}{r_*} = \left(\frac{t}{t_*} \right)^{1/2} = \left(\frac{C}{C_*} \right)^{-1/3}.$$

Hence, we can form invariant combinations. From the two independent variables, cross-multiplying, we obtain

$$\eta_1 = rt^{-1/2} = r_* t_*^{-1/2} \quad \text{and} \quad \eta_2 = t^{3/2} C = t_*^{3/2} C_*. \tag{8}$$

For the solution to have the same invariance properties as the partial differential equation and the boundary and initial conditions then the solution must also be invariant under the stretching transformation (7). This can be achieved by functionally relating the two invariant combinations in (8) by writing $\eta_2 = f(\eta_1)$. This gives

$$C(r, t) = t^{-3/2} f(\eta) \quad \text{where } \eta = rt^{-1/2}. \tag{9}$$

Reduction of the PDE to an ODE

In (9) we have a similarity transformation which is guaranteed to reduce the partial differential equation (1) to an ordinary differential equation in the similarity variable η. The following example does this.

Example 2: *Use the chain rule to reduce the PDE to an ODE.*

Solution: *Using the product rule and then the chain rule, we obtain*

$$\frac{\partial C}{\partial t} = \frac{\partial}{\partial t}(t^{-3/2}f(\eta))$$

$$= \frac{-3}{2}t^{-5/2}f(\eta) + t^{-3/2}\frac{df}{d\eta}\frac{\partial\eta}{\partial t}$$

$$= \frac{-3}{2}t^{-5/2}f + \left[t^{-3/2}\frac{df}{d\eta} \times \left(-\frac{1}{2}rt^{-3/2}\right)\right].$$

Substituting $r = \eta t^{1/2}$, *we obtain*

$$\frac{\partial C}{\partial t} = -\frac{1}{2}t^{-5/2}\left(3f + \eta\frac{df}{d\eta}\right). \tag{10}$$

Using the chain rule on the r*-derivative gives*

$$\frac{\partial C}{\partial r} = \frac{\partial}{\partial r}\left(t^{-3/2}f(\eta)\right) = t^{-3/2}\frac{df}{d\eta} \times \frac{\partial\eta}{\partial r} = t^{-3/2}\frac{df}{d\eta} \times t^{-1/2} = t^{-2}\frac{df}{d\eta}.$$

Multiplying by r^2, *and using* $r = t^{1/2}\eta$ *to eliminate* r, *we obtain*

$$r^2\frac{\partial C}{\partial r} = t^{-2} \times t\eta^2\frac{df}{d\eta}.$$

Using the chain rule again, gives

$$\frac{\partial}{\partial r}\left(r^2\frac{\partial C}{\partial r}\right) = t^{-3/2}\frac{d}{d\eta}\left(\eta^2\frac{\partial f}{\partial\eta}\right).$$

Hence, dividing by r^2 *and again letting* $r = t^{1/2}\eta$,

$$\frac{1}{r^2}\frac{\partial}{\partial r}\left(r^2\frac{\partial C}{\partial r}\right) = t^{-5/2}\frac{1}{\eta^2}\frac{d}{d\eta}\left(\eta^2\frac{df}{d\eta}\right). \tag{11}$$

Equating (10) and (11), dividing through by $t^{-5/2}$ *and multiplying through by* η^2, *the partial differential equation (1) reduces to the ordinary differential equation*

$$-\frac{1}{2}\left(3\eta^2f + \eta^3\frac{df}{d\eta}\right) = \frac{d}{d\eta}\left(\eta^2\frac{df}{d\eta}\right). \tag{12}$$

Solving the ODE

The solution of this differential equation is fairly involved. The details are left for the problems at the end of this chapter, with an overview given here. The method of solution involves recognising that $(\eta^3 f)' = 3\eta^2 f + \eta^3 f'$ which allows an integration of the differential equation (see exercises, Questions 10 and 11 or Question 12). A family of solutions is

$$f(\eta) = c_2 e^{-\eta^2/4} \tag{13}$$

where c_2 is an arbitrary constant. Using $C(r,t) = t^{-3/2} f(\eta)$, $\eta = rt^{-1/2}$, from (9), the concentration is given by

$$C(r,t) = c_2 t^{-3/2} e^{-r^2/4t}. \tag{14}$$

Condition (13) allows us to evaluate the constant c_2. We can't apply this condition directly, since the solution is singular at $t = 0$. However, by conservation of mass, the total mass $M_0 = (4\pi)^{-1}$ distributed over all space remains constant, so that

$$\int_0^\infty r^2 C(r,t)\, dr = \frac{1}{4\pi},$$

or in terms of the similarity variables

$$\int_0^\infty \eta^2 f(\eta)\, d\eta = \frac{1}{4\pi}$$

which can then be used to find c_2 (see Question 13). This gives $c_2 = (4\pi)^{-3/2}$ and so

$$C(r,t) = \left(\frac{1}{4\pi t}\right)^{3/2} e^{-r^2/4t}. \tag{15}$$

Interpretation

At $r = 0$ we have the concentration proportional to $t^{-3/2}$. This shows that the concentration becomes unbounded as $t \to 0$. Of course, in practice, the concentration cannot be infinite, but this model reflects

that the pollutant is very concentrated initially compared to what it is
later on. For large times we see that the concentration tends to zero.

Now let us consider the spatial variation of concentration at different times. For very small t the argument of the exponential is large
and negative provided r is not small. Hence, except for very close to
the origin, the concentration is very close to zero. For large times the
argument of the exponential is small hence the concentration does not
depend strongly on r for large t (i.e. the concentration is approximately
uniform). In Figure 3.3.2 we sketch the concentration distributions at
different times.

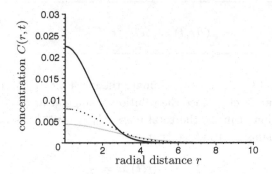

Fig. 3.3.2. Distribution of concentrations at various times ($t = 1$, black; $t = 2$,
dots; and $t = 3$ grey) due to an injection of mass concentrated at $r = 0$ at
time $t = 0$.

Now, let us describe what is happening in this problem. Initially,
an "injection" of mass is put into the fluid. Because the mass is concentrated at a single point the concentration is infinite. Immediately
diffusion begins. This tends to smooth out the initial distribution of
concentration as mass diffuses away from the origin. Shortly after $t = 0$
the concentration is sharply peaked at the origin. Subsequently as the
pollutant diffuses away from the source, the concentration distribution
flattens out.

3.4 Solving the water filtration case study

In this section we apply the stretching transformation approach developed in the previous two sections. This is used to derive an appropriate
similarity solution for the case study problem discussed in Section 3.1.

Review of the problem

In Section 3.1 we developed governing equations for the process of reverse osmosis near a semi-permeable boundary (see Figure 3.1.2). In this problem a saline solution flows along a semi-permeable boundary. Pure water passes through the boundary since salt molecules are prevented from passing through. Thus the salt concentration builds up on the boundary. Our task is to predict the long term salt concentration on the boundary as a function of distance along the semi-permeable boundary.

The equilibrium concentration satisfies an advection-diffusion equation (see Section 1.5). If we define the *excess* equilibrium concentration $c(x,y)$ by

$$c(x,y) = C(x,y) - c_0 \tag{1}$$

where c_0 is the salt concentration at the entrance to the filter, then we obtained, in Section 3.1, the approximate governing equations for the excess concentration

$$y\frac{\partial c}{\partial x} = \alpha\frac{\partial^2 c}{\partial y^2}, \qquad \alpha = \frac{Dv_0}{h} \tag{2}$$

where v_0 is the velocity of the fluid at a distance h above the semi-permeable plate. The boundary conditions were

$$c(0,y) = 0, \qquad c(x,\infty) = 0. \tag{3}$$

We also have the boundary condition on the membrane surface

$$-D\frac{\partial c}{\partial y}(x,0) = qc_0. \tag{4}$$

Using stretching transformations

We can now apply the method of stretching transformations developed in the previous two sections to find a transformation which reduces the partial differential equation to an ordinary differential equation. A general stretching transformation of the variables x, y and c is given by

$$x = e^a x_*, \quad y = e^b y_*, \quad c = e^\gamma c_*.$$

If we substitute this into the partial differential equation and boundary conditions and require invariance we obtain (see problems, Question 15) the similarity transformation

$$c(x, y) = x^{1/3} f(\eta), \qquad \eta = y x^{-1/3}. \tag{5}$$

Substituting (5) into the partial differential equation (2) then we obtain the ordinary differential equation (see problems, Question 15)

$$\frac{\eta}{3}\left(f - \eta \frac{df}{d\eta}\right) = \alpha \frac{d^2 f}{d\eta^2}. \tag{6}$$

This is a linear second-order differential equation.

Solving the ODE

The solution of this linear second-order differential equation and application of the boundary conditions is somewhat lengthy so only an overview is given here, with the details explored in the problems at the end of this chapter. By inspection, we can see that $f(\eta) = \eta$ is one solution. To obtain the general solution the method of reduction of order can be used, where we assume a second solution of the form $f_2(\eta) = \eta g(\eta)$. The details are carried out in the problems (see Question 16 and Question 17). Alternatively, we may be able to use a computer algebra package, such as Maple. The general solution of the differential equation, using the boundary condition $f(\infty) = 0$ is

$$f(\eta) = C_2 \eta \int_\eta^\infty \frac{e^{-\eta_1^3/(9\alpha)}}{\eta_1^2} \, d\eta_1 \tag{7}$$

and applying the boundary condition at $y = 0$ it is possible to obtain, after some algebra (see problems Question 17), the solution on $y = 0$

expressed as

$$c(x,0) = \frac{3\alpha q c_0}{D I_1} x^{1/3}, \qquad I_1 = (9\alpha)^{2/3} \int_0^\infty v e^{-v^3}\, dv. \qquad (8)$$

Discussion of the solution

The definite integral I_1 may be obtained approximately by numerical integration. Using numerical integration (e.g. Simpson's rule, or Maple) its value is given, to 2 decimal places, as

$$I_1 \simeq 1.09(9\alpha)^{2/3}. \qquad (9)$$

Note that $\alpha = Dh/v_0$, where D is the diffusivity of salt in water, v_0 is the horizontal velocity measured at a distance h from the semi-permeable boundary. Using this, (1) and (9), the concentration $C(x,0)$ of salt on the semi-permeable boundary is given by

$$C(x,0) = c_0 + 0.64\frac{q c_0}{D}\left(\frac{Dh}{v_0}\right)^{1/3} x^{1/3}. \qquad (10)$$

The concentration increases with distance along the plate, but this increase occurs slowly due to the presence of the $x^{1/3}$ factor. With the parameter values chosen, in Figure 3.4.1, the proportional increase is quite modest (only 0.2% in 200 cm), however, this increases if we increase the flow rate q through the semi-permeable membrane.

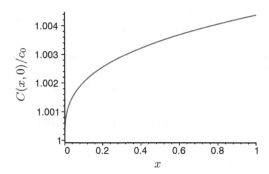

Fig. 3.4.1. Concentration ratio C/c_0 along the plate. The parameter values $h = 10^{-3}$, $D = 10^{-9}$, $v_0 = 10^{-3}$ and $q = 10^{-3}$ have been used.

Further reading

This case study was based on a problem considered in Probstein (1989). A discussion of a practical device for water filtration using this technology is given in James et al. (1993) and also Merten (1966).

   Some further discussion of the mathematics of filtration can be found in Probstein (1989) and Lightfoot (1974). A good introduction to the theory of stretching transformations and similarity solutions can be found in Dresner (1983). For those wishing to explore how the theory can be extended to more general transformations (the method of Lie symmetries), Logan (1987) gives an introduction, as does Hill (1992). A more advanced exposition is given in Bluman and Kumei (1989).

3.5 Problems for Chapter 3

1.   *Given the stretching transformation*

$$x = e^a x_*, \quad t = e^b t_*, \quad u = e^\gamma u_*,$$

express $\dfrac{\partial^3 u}{\partial x \partial t^2}$ *and* $\dfrac{\partial u}{\partial x}\dfrac{\partial u}{\partial t}$ *in the starred variables.*

2.   *Identify which of the following functions are not invariant under the stretching transformation* $x = e^a x_*, \quad t = e^{2a} t_*, \quad u = e^a u_*.$

   (a) $u = xt^{-1}$ (b) $u = \sqrt{t}e^{x^2/t}$ (c) $u = t\sin x^t.$

3.   *Consider a general similarity transformation of the form,*

$$u(x,t) = t^\gamma f\left(\frac{x}{\sqrt{t}}\right)$$

where f is some function and γ is a constant. Substitute this into the partial differential equation

$$\frac{\partial u}{pt} = \alpha \frac{\partial^2 u}{\partial x^2}$$

and show that f satisfies the ordinary differential equation

$$\gamma f(\eta) - \frac{1}{2}\eta f(\eta) = \alpha f''(\eta).$$

4.   Consider the heat conduction problem

$$\frac{\partial u}{\partial t} = \alpha \frac{\partial^2 u}{\partial x^2}$$

with a prescribed heat flux on the boundary $x = 0$,

$$-k \frac{\partial u}{\partial x}(0, t) = 1,$$

and zero temperature initially.

(a) Use the stretching transformation method to find a similarity variable for the problem.

(b) Obtain an ordinary differential equation for f. What are the boundary conditions for f?

5.   Consider 1-D nonlinear diffusion in a semi-infinite tube where the diffusivity is proportional to the concentration. The appropriate diffusion equation is the nonlinear PDE

$$\frac{\partial C}{\partial t} = \frac{\partial}{\partial x}\left(C \frac{\partial C}{\partial x}\right)$$

Suppose, also, that $C(0, t) = 1$ and the initial concentration is zero.

(a) Find the group of stretching transformations that leave the partial differential equation, boundary condition and initial condition simultaneously invariant.

(b) Hence find an appropriate similarity substitution.

(c) Hence reduce the partial differential equation to an ordinary differential equation. What are the boundary conditions for the ordinary differential equation?

6.   Consider the partial differential equation

$$u \frac{\partial u}{\partial t} = \frac{\partial^2 u}{\partial x^2}.$$

with the boundary and initial conditions

$$u(x, 0) = 0, \qquad u(\infty, t) = 0, \qquad \frac{\partial u}{\partial x}(0, t) = -1.$$

(a) Using the method of stretching transformations show that

$$u(x, t) = t^{1/3} f(xt^{-1/3}).$$

(b) Hence obtain the ordinary differential equation satisfied by f and write down the boundary conditions satisfied by f.

7. Consider the partial differential equation, for a semi-infinite region $(0, \infty)$,

$$u\frac{\partial u}{\partial t} + \left(\frac{\partial u}{\partial x}\right)^2 = 0$$

with temperature at $x = 0$ given by $u_0\sqrt{t}$.

(a) Using the method of stretching transformations show that

$$u(x, t) = \sqrt{t} f(x/\sqrt{t}).$$

(b) Hence obtain the ordinary differential equation satisfied by f.

8. The partial differential equation

$$\frac{\partial C}{\partial t} = \frac{\partial}{\partial x}\left(\sqrt{x}\frac{\partial C}{\partial x}\right)$$

arises from diffusion with the diffusivity proportional to the square root of the distance from the origin. The initial and boundary conditions are

$$C(0, t) = 0, \qquad C(\infty, t) = 1, \qquad C(x, 0) = 1.$$

(a) Use the stretching transformation method to show $C(x, t) = f(\eta)$ where $\eta = x/t^{2/3}$.

(b) Hence show that f satisfies

$$\frac{d}{d\eta}(\sqrt{\eta}f'(\eta)) + \frac{2}{3}\eta f'(\eta) = 0$$

9. For the previous question, obtain a solution in terms of an integral. [Hint: Try the substitution $v = \sqrt{\eta}$.]

10. Consider the differential equation

$$-\frac{1}{2}\left(3\eta^2 f + \eta^3\frac{df}{d\eta}\right) = \frac{d}{d\eta}\left(\eta^2\frac{df}{d\eta}\right) \qquad (1)$$

which arose in the point source problem in Section 3.3.

Using $(\eta^3 f)' = 3\eta^2 f + \eta^3 f'$, solve the differential equation to obtain the general solution

$$f(\eta) = -c_1 e^{-\eta^2/4}\int_a^\eta \frac{e^{\eta_1^2/4}}{\eta_1^2}\, d\eta_1 + c_2 e^{-\eta^2/4}$$

where c_1 and c_2 are arbitrary constants and a is an arbitrary positive number.

11. Consider the general solution to the point source problem obtained in Question 10, which may be written as

$$f(\eta) = c_1 f_1(\eta) + c_2 f_2(\eta)$$

where

$$f_1(\eta) = -e^{-\eta^2/4} \int_a^\eta \frac{e^{\eta_1^2/4}}{\eta_1^2} \, d\eta_1, \qquad f_2(\eta) = e^{-\eta^2/4}.$$

(a) L'Hôpital's rule states:

$$\lim_{x\to\infty} \frac{g(x)}{h(x)} = \lim_{x\to\infty} \frac{g'(x)}{h'(x)}$$

if both

$$\lim_{x\to\infty} g(x) = \infty \quad and \quad \lim_{x\to\infty} h(x) = \infty.$$

Use this to show that $f(\infty) = 0$ for all values of c_1 and c_2 and hence this boundary condition cannot be used to determine either arbitrary constant.

(b) The integral condition

$$\int_0^\infty \eta^2 f(\eta) \, d\eta = \frac{1}{4\pi}$$

sets some conditions on the rate of decay of the function f as $\eta \to \infty$; for the integral to converge then the integrand must decay more rapidly than η^{-1} as $\eta \to \infty$ for the integral to converge, which requires that f must decay faster than η^{-3}.

By considering the limit as $\eta \to \infty$ of $\eta^3 f_1$ and $\eta^3 f_2$ deduce that $c_1 = 0$. (You can do this by plotting the appropriate functions, or more rigorously by repeatedly using L'Hopital's rule.)

12. Consider the differential equation

$$-\frac{1}{2} \left(\eta^3 f\right)' = \left(\eta^2 f'\right)'$$

which arises in the solution for the point source in Section 3.3.

(a) Integrate the differential equation to obtain

$$-\frac{1}{2}\eta^3 f = \eta^2 f' + c_1$$

where c_1 is an arbitrary constant.

(b) Given that f decays faster than η^{-3} deduce that $c_1 = 0$. Hence solve the resulting differential equation to obtain

$$f(\eta) = c_2 e^{-\eta^2/4}$$

where c_2 is an arbitrary constant.

13.   For the point source problem in Section 3.3, a family of possible solutions for f is (see Question 11 or Question 12)

$$f(\eta) = c_2 e^{-\eta^2/4}.$$

Apply the integral condition

$$\int_0^\infty \eta^2 f(\eta) \, d\eta = \frac{1}{4\pi}$$

to obtain the arbitrary constant c_2.

14.   Consider 1-D diffusion on $-\infty < x < \infty$ with diffusivity proportional to concentration,

$$\frac{\partial C}{\partial t} = \frac{\partial}{\partial x} \left(C \frac{\partial C}{\partial x} \right).$$

A source of unit mass is injected at $x = 0$ at $t = 0$.

(a) Given unit cross-sectional area, explain why

$$\int_{-\infty}^\infty C(x,t) \, dx = 1.$$

Write down the boundary conditions.

(b) Use the stretching transformation method to find a suitable transformation.

(c) Hence find an ordinary differential equation (but do not solve).

15.   For a concentration $c(x, y)$, consider the partial differential equation

$$y \frac{\partial c}{\partial x} = \alpha \frac{\partial^2 c}{\partial y^2}$$

where α is a constant, and the boundary conditions are

$$c(0, y) = 0, \quad c(x, \infty) = 0, \quad \frac{\partial c}{\partial y}(x, 0) = -\frac{q_0 c_0}{D}.$$

(a) Use the method of stretching transformations to find a similarity transformation which reduces the PDE to an ODE.

(b) Hence find the ODE obtained by applying the similarity transformation.

(c) Express the boundary conditions for the PDE in terms of the boundary conditions for the ODE.

16. In Section 3.4 we obtained the differential equation

$$\frac{\eta}{3}\left(f - \eta\frac{df}{d\eta}\right) = \alpha\frac{d^2 f}{d\eta^2}. \tag{2}$$

(a) **Verify** that $f(\eta) = \eta$ is a solution and look for a second solution by substituting $f(\eta) = \eta g(\eta)$ obtaining a first-order differential equation for g'. Hence obtain the general solution

$$f(\eta) = C_1\eta + C_2\eta\int_\eta^\infty \frac{e^{-\eta_1^3/(9\alpha)}}{\eta_1^2}\, d\eta_1$$

where C_1 and C_2 are arbitrary constants.

17. Given the general solution (7) for the case study problem in Section 3.4

$$f(\eta) = C_2\left[-e^{-\eta^3/(9\alpha)} - \frac{\eta}{3\alpha}\int_\eta^\infty \eta e^{-\eta_1^3/(9\alpha)}\, d\eta_1\right]$$

apply the boundary condition

$$f'(0) = \frac{-q_0}{D}$$

to obtain

$$C_2 = \frac{3^{-1/3}\alpha^{1/3}qc_0}{DI_1} \qquad \text{where} \quad I_1 = \int_0^\infty ve^{-v^3}\, dv.$$

18. Given

$$\lim_{\eta\to\infty}\int_\eta^\infty e^{-\eta_1^3/(9\alpha)}\, d\eta_1 = 0$$

use L'Hôpital's rule (see Question 10) to prove that $f(\infty) = 0$ implies that $C_1 = 0$ in (7) (alternatively, you can prove the above limit using a squeeze principal, i.e. find simpler functions which bound the given function above and below).

19. Use integration by parts to prove

$$\int_\eta^\infty \eta_1^{-2}e^{-\eta_1^3/(9\alpha)}\, d\eta_1 = -\eta^{-1}e^{-\eta^3/(9\alpha)} - \frac{\eta}{3\alpha}\int_\eta^\infty \eta e^{-\eta_1^3/(9\alpha)}\, d\eta_1.$$

4

Case Study: Laser Drilling

In this chapter we develop a mathematical model to calculate the drilling speed of a laser through a thick sheet of metal. We can easily find an expression for the drilling speed if we neglect heat conduction into the metal. To include heat conduction, we introduce the method of regular perturbations. This allows us to develop equations for a "correction" due to a small term (in this case due to conduction) which otherwise makes the the equations too difficult to solve.

4.1 Introduction to the case study problem

There is considerable interest in industry in the use of high power lasers and electron beams for both cutting and welding of sheet metal. The essential idea is to focus a large amount of energy onto a small area of the surface of the metal. This intense heating causes vaporisation of the metal, forming a hole. The following case study is based on a chapter of Andrews and McLone (1976).

The problem

There are a number of mathematical questions associated with this problem. The one we try to answer is that of determining how rapidly the hole is formed for a given value of the power per unit area of the irradiating beam. This is a heat conduction problem with a phase change (solid to gas) and a moving boundary.

As energy from the laser is absorbed by the surface, the temperature of

112

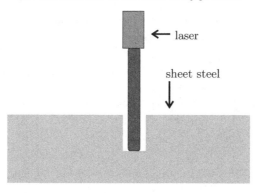

Fig. 4.1.1. Schematic diagram of laser drilling. A laser of power Q_0 drills through metal by vaporising the metal.

the metal rises and heat is conducted through the metal. The temperature cannot rise indefinitely. At a certain temperature (the vaporisation temperature, u_v) energy is absorbed by the material as it changes from solid to vapour (we will ignore the liquid phase — the heat supplied by the laser is so intense that the metal appears to boil at once). We will denote the amount of heat (per unit mass) required for the metal to vaporise by λ, the specific latent heat of vaporisation.

We will further ignore any hydrodynamic effects, assuming the metal vapour is drawn off before it can turn back to liquid, and we ignore the thermal expansion of the metal.

There are several calculations to be done. One involves calculating the time for the boundary $x = 0$ to heat to the vaporisation temperature. It turns out that this happens very quickly. Another calculation justifies that it is reasonable to neglect heat flow in the x and y directions. These calculations are described in detail in Andrews and McLone (1976).

A 1-D model

Let us assume that the moving boundary advances a distance δs in a time interval t to $t + \delta t$ (see Figure 4.1.2). In doing so, some energy from the laser is used in changing the metal from solid to vapour. We will neglect any heat energy used to raise the temperature of the mass $\rho A \delta s$ of metal to the vaporisation temperature, expecting this to be small compared to the heat required to vaporise a mass $\rho A \delta s$ of metal. Thus the heat energy not used to vaporize the metal is conducted away.

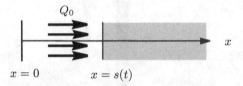

$$Q_0$$

$$x = 0 \qquad x = s(t)$$

Fig. 4.1.2. A one-dimensional, semi-infinite model of the hole formation process.

We do an overall heat balance. In a time δt conservation of energy requires

$$\left\{\begin{array}{c}\text{latent heat used}\\ \text{to vaporize}\\ \text{material}\end{array}\right\} + \left\{\begin{array}{c}\text{amount of heat}\\ \text{conducted}\\ \text{away}\end{array}\right\} = \left\{\begin{array}{c}\text{heat supplied}\\ \text{by laser}\\ \text{to surface}\end{array}\right\}. \qquad (1)$$

Let λ be the latent heat per unit mass of the metal required to vaporise the metal. Then

$$\left\{\begin{array}{c}\text{latent heat required}\\ \text{to vaporize material}\end{array}\right\} \simeq \lambda \rho A \,\delta s,$$

where ρ is the density of the material and A the area on which the laser is focused. The amount of heat conducted away is given, in terms of the heat flux J, by

$$\left\{\begin{array}{c}\text{amount of heat}\\ \text{conducted}\\ \text{away}\end{array}\right\} = AJ(s(t + \delta t), t)\delta t.$$

If the power supplied by the laser is Q_0 watts then

$$\left\{\begin{array}{c}\text{heat supplied}\\ \text{by laser}\\ \text{to surface}\end{array}\right\} \simeq Q_0 \,\delta t.$$

Putting this together, we obtain

$$\lambda \rho A \delta s + AJ(s(t + \delta t), t)\delta t = Q_0 \delta t.$$

Dividing through by δt and A and then letting $\delta t \to 0$ gives

$$\rho \lambda \frac{ds}{dt} = \frac{Q_0}{A} - J(s(t), t).$$

Using Fourier's law of heat conduction, we have

$$\rho\lambda\frac{ds}{dt} = \frac{Q_0}{A} + k\frac{\partial u}{\partial x}(s(t), t). \qquad (2)$$

This is an equation for the drilling speed ds/dt.

We also need equations for the temperature $u(x, t)$ in the metal. Now let us write down the complete set of governing equations for the simple one-dimensional, semi-infinite model of the hole forming process. We assume that the temperature in the metal satisfies the classical heat equation

$$\frac{\partial u}{\partial t} = \alpha\frac{\partial^2 u}{\partial x^2}, \qquad \text{for } x > s(t). \qquad (3)$$

Let us assume the initial temperature is zero but the surface of the boundary has been pre-heated to the vaporisation temperature. The initial temperature distribution is

$$u(x, 0) = 0, \qquad (4)$$

and the boundary conditions are

$$u(s(t), t) = u_v \qquad (5)$$

and

$$u(\infty, t) \to 0. \qquad (6)$$

With this set of governing equations, bearing in mind the similarity to those used in Chapter 2, we might be tempted to seek a Boltzmann similarity solution of the form $u = f(\phi)$ where $\phi = x/\sqrt{\alpha t}$. Unfortunately the presence of the Q_0/A term in the modified Stefan condition means that the classical similarity approach as used in Chapter 2 will not work here. Thus it appears that we might eventually have to resort to a numerical solution. First, however, we can try to simplify the problem and find an approximate solution to the equations.

A first estimate for the drilling speed

One possible simplification is to neglect heat conduction. The reason for trying to neglect heat conduction is an expectation that most of the heat is used up in vaporising the metal. If we can neglect the conduction of heat in the metal we can easily obtain an expression for the speed of drilling. Setting $k = 0$ in (2) gives a drilling speed estimate, v_0, as

$$v_0 = \frac{ds}{dt} = \frac{Q_0}{\rho \lambda A}. \qquad (7)$$

Let us consider a 1 kW laser focused on an area 1 mm^2 drilling through a thick sheet of aluminium. For aluminium $\rho = 2.7 \times 10^3 \, \mathrm{kg \, m^{-3}}$, $\lambda = 1.08 \times 10^7 \, \mathrm{J \, kg^{-1}}$. This estimate gives a drilling speed of $v_0 = 34 \, \mathrm{mm/s}$.

Scaled equations

We shall make the governing equations for the 1-D conduction model dimensionless so that we can see when terms in these equation are small, compared to other terms. First we need to choose suitable scales for the variables.

The variables we have to scale are

$$x, \quad t, \quad s \quad \text{and} \quad u.$$

In the problem the constants available to scale them with are

$$u_v, \quad k, \quad \alpha, \quad \rho, \quad \lambda, \quad Q_0 \quad \text{and} \quad A.$$

Obviously, we can scale the temperature u with respect to the vaporisation temperature u_v. There are no obvious length scales or time scales, however, so we must construct them from the available variables. It is natural to use the previous estimate for the drilling speed v_0.

Let us introduce the symbols x_0 and t_0 for the yet to be determined length and time scales. Now, since v_0 is a velocity, a length scale is given by $x_0 = v_0 t_0$. A time scale for heat conduction is $t_0 = x_0^2/\alpha$. Substituting t_0 into the expression for x_0 we obtain

$$x_0 = \frac{\alpha}{v_0} \quad \text{and} \quad t_0 = \frac{\alpha}{v_0^2}.$$

Note that these have the correct dimensions of length and time respectively.

We now define the dimensionless variables U, S, X and T corresponding to u, s, x and t by

$$U = \frac{u}{u_v}, \quad S = \frac{s}{x_0} = \frac{v_0}{\alpha}s, \quad X = \frac{x}{x_0} = \frac{v_0}{\alpha}x \quad \text{and} \quad T = \frac{t}{t_0} = \frac{v_0^2}{\alpha}t.$$

Substituting this into the governing equations (2)–(6) we obtain (see exercises, Question 2) the following dimensionless equations:

$$\frac{dS}{dT} = 1 + \epsilon\frac{\partial U}{\partial X} \quad \text{on } X = S(T),$$

with the dimensionless governing equation for temperature

$$\frac{\partial U}{\partial T} = \frac{\partial^2 U}{\partial X^2}$$

and with the initial and boundary conditions

$$U(X, 0) = 0, \quad U(S(T), T) = 1, \quad U(\infty, T) = 0.$$

The dimensionless parameter ϵ is given by

$$\epsilon = \frac{ku_v}{\rho\alpha\lambda} = \frac{cu_v}{\lambda}. \tag{8}$$

Physically the parameter ϵ represents the ratio of the heat used to vaporise the metal to the heat used in raising the temperature to the vaporisation temperature. Looking over the table below, we see that for most common metals ϵ may be regarded as a **small** parameter.

Table 4.1.1. *Thermal properties for some common materials. Here c is the specific heat, u_v is the vaporisation temperature and λ is the specific latent heat of vaporisation.*

Material	u_v °C	λ $J\,kg^{-1}$	c $J\,kg^{-1}\,°C$	ϵ
Aluminium	2 727	1.080×10^7	913	0.23
Copper	2 499	4.770×10^7	385	0.20
Gold	2 788	1.740×10^7	132	0.21
Iron	12 170	6.070×10^7	106	0.21
Lead	1 722	0.861×10^7	126	0.25
Nickel	2 739	6.361×10^7	460	0.20
Zinc	885	1.780×10^7	385	0.19

So that we can investigate the effect of conduction on the drilling speed
we need to develop an approximate solution which somehow takes advantage of the parameter ϵ being small. To do this we introduce the
method of perturbation expansions.

Since $\{T, X, U, \epsilon\}$ is a complete set of dimensionless variables and parameters for this problem, the solution $U(X, T)$ must also be a function
of ϵ. We consider a power series in ϵ. Essentially we look for a solution
$U(X, T)$ and $S(T)$ of the form

$$U = U_0 + \epsilon U_1 + \epsilon^2 U_2 + \ldots, \qquad S = S_0 + \epsilon S_2 + \epsilon^2 S_2 + \ldots,$$

where U_0 and S_0 correspond to the approximate solution when $\epsilon =
0$. The terms U_1 and S_1 represent a correction to the original U_0, S_0
solutions. The terms U_2, S_2 are higher order corrections.

To develop familiarity with the technique we will first look at some
simpler problems. First, in Section 4.2, we will look at a simple ordinary
differential equation where one of the terms in the boundary conditions
is multiplied by a small parameter. Second, in Section 4.3 we will look
at a two-dimensional, steady-state heat conduction problem where the
small parameter occurs inside the boundary condition.

4.2 Method of perturbations

Here we develop a procedure for obtaining a sequence of approximations
where each successive term is a small correction to the previously obtained term. It is known as the method of perturbations. We introduce
a simple example problem on which to illustrate the method.

Example problem

The dimensionless equations for an equilibrium heat conduction problem
are

$$\frac{d^2U}{dX^2} - \epsilon U^4 = 0 \qquad (1)$$

with boundary conditions

$$U(0) = 1 \quad \text{and} \quad U(1) = 0 \tag{2}$$

where $\epsilon \ll 1$. In this problem there is a heat source term which arises from radiation (which contributes the U^4 term).

For small ϵ we want to develop a sequence of approximate solutions for the dimensionless temperature U where the zeroth-order term corresponds to $\epsilon = 0$ and where we can obtain a first-order correction to this, and then a second-order correction and so on.

The general procedure is to assume an expansion of the form

$$U = U_0 + \epsilon U_1 + \epsilon^2 U_2 + \ldots \tag{3}$$

for the dependent variable, U. Here U_0, U_1, U_2 are functions to be determined, where U_0 is the principal term (or zeroth-order term), U_1 is a correction to the principal term, and U_2 is a further correction, and so on.

Approximate equations

We now substitute this expansion into the governing equations and retain terms up to ϵ^2. The details are shown in the following example.

Example 1: *Find the differential equations for U_0, U_1 and U_2.*

Solution: *Substituting (3) into the differential equation (1) gives*

$$\frac{d^2 U_0}{dX^2} + \epsilon \frac{d^2 U_1}{dX^2} + \epsilon^2 \frac{d^2 U_2}{dX^2} = \epsilon(U_0 + \epsilon U_1 + \ldots)^4.$$

Since we are only retaining terms of order ϵ^2† then, in the term ϵU^4, we only need to retain terms of order ϵ^1 within the expansion for U^4. Expanding $(U_0 + \epsilon U_1 + \ldots)^4$ we obtain

$$\epsilon(U_0 + \epsilon U_1 + \ldots)^4 = \epsilon(U_0^2 + 2\epsilon U_0 U_1 + \ldots)^2 = \epsilon U_0^4 + 4\epsilon^2 U_0^3 U_1 + \ldots.$$

---

† A notation $O(\epsilon^3)$ is often adopted to represent all those terms involving ϵ^3 and higher powers that are neglected.

The differential equation now becomes

$$\frac{d^2U_0}{dX^2} + \epsilon\frac{d^2U_1}{dX^2} + \epsilon^2\frac{d^2U_2}{dX^2} = \epsilon U_0^4 + 4\epsilon^2 U_0^3 U_1 + \dots.$$

Any terms involving ϵ^3 or smaller have been neglected.

We can now extract equations for each of the terms U_0, U_1 and U_2 by equating coefficients of powers of ϵ. This is called "collecting terms of like order". The coefficient of ϵ^0 gives

$$\frac{d^2U_0}{dX^2} = 0. \tag{4}$$

Similarly, the coefficient of ϵ^1 gives

$$\frac{d^2U_1}{dX^2} = U_0^4 \tag{5}$$

and, for ϵ^2, we get

$$\frac{d^2U_2}{dX^2} = 4U_0^3 U_1. \tag{6}$$

We also have to find appropriate sets of boundary conditions for the functions U_0, U_1 and U_2. This is done in the following example.

Example 2: *Find the appropriate boundary conditions for U_0, U_1 and U_2.*

Solution: *Substituting (3) into the boundary conditions (2), we obtain*

$$U_0(0) + \epsilon U_1(0) + \epsilon^2 U_2(0) + \dots = 1,$$
$$U_0(1) + \epsilon U_1(1) + \epsilon^2 U_2(1) + \dots = 0.$$

We equate coefficients of powers of ϵ. Thus, the zeroth-order, first-order and second-order boundary conditions are

$$\begin{aligned}
U_0(0) = 1, \quad U_1(0) = 0, \quad U_2(0) = 0, \\
U_0(1) = 0, \quad U_1(1) = 0, \quad U_2(1) = 0.
\end{aligned} \tag{7}$$

Note the first-order and second-order terms are zero since there are no order ϵ or order ϵ^2 terms on the right-hand side. If the boundary conditions are linear and do not involve ϵ then we expect the non-homogeneous part of the boundary conditions to appear in the U_0 terms where the higher-order terms will be homogeneous.

The solutions

Now we can solve for the various approximations. The zeroth-order equations are (4) with the boundary conditions for U_0 from (7)

$$\frac{d^2 U_0}{dX^2} = 0, \qquad U_0(0) = 1, \quad U(1) = 0. \tag{8}$$

The solution is given in the following example.

Example 3: *Solve the boundary value problem (8).*

Solution: *The solution of this very simple differential equation can be obtained by integrating both sides with respect to X twice, yielding*

$$U_0(X) = c_1 X + c_2$$

where c_1 and c_2 are two arbitrary constants. Applying the two boundary conditions for U_0 gives the equations $1 = c_2$ and $0 = c_1 + c_2$ and hence $c_1 = -1$ and $c_2 = 1$. We thus obtain the solution

$$U_0(X) = 1 - X \tag{9}$$

for the zeroth-order term of the perturbation.

Using the solution for U_0, and substituting into (5) the equations for the first-order correction (with appropriate boundary conditions from (7)) are

$$\frac{d^2 U_1}{dX^2} = U_0^4 = (1 - X)^4, \qquad U_1(0) = 0, \quad U_1(1) = 0. \tag{10}$$

Note that the first-order correction term U_1 depends on knowing the solution for the zeroth-order term U_0.

Example 4: *Solve the boundary value problem (10).*

Solution: *Again the differential equation is of such a simple type that we can integrate both sides twice with respect to X. We obtain*

$$U_1(X) = \frac{1}{30}(1 - X)^6 + c_3 X + c_4$$

where c_3 and c_4 are two arbitrary constants. Applying the two boundary conditions $U_1(0) = 0$ and $U_1(1) = 0$ gives $(1/30) + c_4 = 0$ and $c_3 + c_4 = 0$. Hence $c_4 = -1/30$ and $c_3 = 1/30$. Thus

$$U_1(X) = \frac{1}{30}(1 - X)^6 - \frac{1}{30}(1 - X).$$

The first-order perturbation approximation is given by

$$U(X) = U_0(X) + \epsilon U_1(X) + O(\epsilon^2).$$

We also call this the perturbation approximation correct to $O(\epsilon)$. Substituting for U_0 and U_1 from above we obtain

$$U(X) = (1 - X) + \epsilon \left\{ \frac{1}{30}(1 - X)^6 - \frac{1}{30}(1 - X) \right\} + O(\epsilon^2).$$

The graphs of the two perturbation approximations U_0 and $U_0 + \epsilon U_1$ are plotted in Figure 4.2.1, for $\epsilon = 3$. There is not much difference between the two solutions. For smaller values of ϵ the solutions are even closer (normally $\epsilon < 1$).

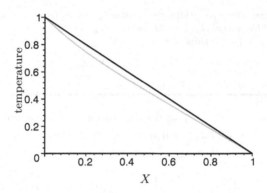

Fig. 4.2.1. Graphs of the two perturbation approximations, U_0 (black line) and $U_0 + \epsilon U_1$ (grey line) plotted against X, for $\epsilon = 3$.

Higher-order approximations

We can continue this process to whatever order is required. However, the work involved usually becomes greater with each order of approximation. It can be shown that the second-order term satisfies

$$\frac{d^2 U_2}{dX^2} = 4U_0^3 U_1 = \frac{2}{15} \left((1 - X)^9 - (1 - X)^4 \right)$$

with the boundary conditions $U_2(0) = 0$ and $U_2(1) = 0$. The solution is straight-forward and can be done by hand (or by using a computer algebra package such as Maple).

Further discussion

There are some subtle difficulties with this approach to finding approximate solutions, for some problems. These are explored in the exercises. We describe two such difficulties briefly.

One such difficulty is for initial-value problems, where the independent variable (usually time) ranges from 0 to ∞. It can happen that the term multiplied by the small parameter can become large so that it is no longer reasonable to neglect that term compared to others in the equation. Thus the perturbation approximation may be valid only for small times (see exercises, Question 9, for an example).

Another complication arises when the small parameter multiplies the highest derivative in the differential equation. These are called *singular perturbations* . The zeroth-order solution cannot then satisfy all the boundary conditions. This is because the order of the differential equation for the zeroth-order term is less than that for the original differential equation. The perturbation approximation will thus not be valid over the entire domain of the problem. (See exercises, Question 10, for an example of a singular perturbation problem.) Since the boundary condition must be satisfied on the boundary there will be a region near the boundary where the true solution changes very rapidly from the value on the boundary to the approximate perturbation solution. This region is called a *boundary layer*. Analysis of boundary layers is very important in fluid dynamics and other fields of study.

Methods have been devised to deal with singular perturbation problems. These involve rescaling the independent variable inside the boundary layer and satisfying just the one boundary condition. The 'inner' perturbation approximation is then patched on to the other 'outer' perturbation solution. This is called the method of matched asymptotic expansions. (See any reasonable book on perturbations or advanced mathematical methods for further details and examples, e.g. Logan (1987); Holmes (1995).)

4.3 Boundary perturbations

In this section an approximate solution is found for a problem where a boundary is perturbed from a simpler boundary. This will develop the necessary skills needed to handle the case study problem, since there the moving boundary is perturbed. The basic idea is to use a Taylor expansion. The method is illustrated on an example problem.

Example problem

When electric current runs through metal some of the energy of the drifting charged particles is converted to heat energy. For a wire of perfectly circular cross-section it is not difficult to obtain an expression for the temperature in the wire. The maximum temperature in the wire occurs at the centre of the cross-section. No manufacturing process is perfect, so the cross-section may not be perfectly circular. Our aim is to investigate the effect of a non-circular cross-section on the temperature in the wire.

Consider a wire where the cross-section of the wire has radius given by

$$R = 1 + \epsilon \cos(\theta)$$

where $\epsilon \ll 1$ is a small parameter. The cross-section is shown in Figure 4.3.1. If $\epsilon = 0$ the cross-section would be perfectly circular. The $\epsilon \cos(\theta)$ term represents a small variation from the circular cross-section.

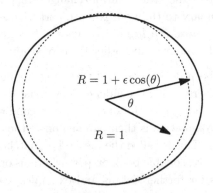

Fig. 4.3.1. A cross-section of an electrical wire which is almost circular.

We suppose the non-dimensional equilibrium temperature of the wire, $U(R, \theta, t)$, satisfies the 2-D Laplace equation with heat source,

$$\nabla^2 U + 1 = 0.$$

In cylindrical polar coordinates (R, θ) this takes the form

$$\frac{1}{R}\frac{\partial}{\partial R}\left(R\frac{\partial U}{\partial R}\right) + \frac{1}{R^2}\frac{\partial^2 U}{\partial \theta^2} = -1. \tag{1}$$

This problem corresponds physically to equilibrium heat conduction with a volumetric heat source within the wire due to electrical resistance.

For boundary conditions let us assume that the temperature at $R = 0$ is finite. On the surface of the wire $R = 1 + \epsilon \cos(\theta)$ we assume that the temperature is held fixed at a temperature $1\,°C$,

$$U(1 + \epsilon\cos(\theta), \theta) = 1. \tag{2}$$

We also have an implicit boundary condition

$$U \text{ is finite} \qquad \text{at } R = 0. \tag{3}$$

To solve this problem we might need to use numerical methods due to the complicated boundary conditions. However, for small values of the parameter ϵ we can find useful approximate solutions using the method of perturbations.

Perturbation of the governing equations

Let us look for a perturbation approximate solution for the temperature involving a zeroth-order term and a first-order correction. This means we assume a solution of the form

$$U = U_0 + \epsilon U_1 + \ldots. \tag{4}$$

The zeroth-order term U_0 will correspond to the problem of a perfectly circular cross-section, corresponding to $\epsilon = 0$. The first-order term represents a first correction to that solution to take account of the non-circular cross-section.

An outline of the procedure is:

- Substitute (4) into the governing equations.
- Collect terms for coefficients of ϵ^0 and ϵ^1.
- Solve the zeroth-order equations.

- Using the zeroth-order solution, solve for the first-order correction.

The following example shows how to do the first two steps.

Example 1: *Substitute the form of the perturbation series (4) into the partial differential equation (1) to obtain the zeroth-order and first-order equations.*

Solution: *Carrying out the substitution we obtain*

$$\frac{1}{R}\frac{\partial}{\partial R}\left(R\frac{\partial}{\partial R}(U_0 + \epsilon U_1 + \ldots)\right) + \frac{1}{R^2}\frac{\partial^2}{\partial \theta^2}(U_0 + \epsilon U_1 + \ldots) = -1.$$

This simplifies to

$$\frac{1}{R}\frac{\partial}{\partial R}\left(R\frac{\partial U_0}{\partial R}\right) + \epsilon\frac{1}{R}\frac{\partial}{\partial R}\left(R\frac{\partial U_1}{\partial R}\right) + \ldots$$

$$+ \frac{1}{R^2}\frac{\partial^2 U_0}{\partial \theta^2} + \epsilon\frac{1}{R^2}\frac{\partial^2 U_1}{\partial \theta^2} + \ldots = -1. \tag{5}$$

Collecting all the coefficients of the ϵ^0 terms gives

$$\frac{1}{R}\frac{\partial}{\partial R}\left(R\frac{\partial U_0}{\partial R}\right) + \frac{1}{R^2}\frac{\partial^2 U_0}{\partial \theta^2} = -1. \tag{6}$$

Similarly, for the coefficients of the ϵ^1 terms,

$$\frac{1}{R}\frac{\partial}{\partial R}\left(R\frac{\partial U_1}{\partial R}\right) + \frac{1}{R^2}\frac{\partial^2 U_1}{\partial \theta^2} = 0. \tag{7}$$

We now deduce the zeroth-order and first-order boundary conditions from (2). The perturbation parameter ϵ appears *inside* the boundary condition. We call this a **boundary perturbation**. If ϵ were identically zero the boundary condition would be a simple one to apply, $U(1, \theta) = 1$. To use the method of perturbations we must first expand U as a Taylor series in the perturbation parameter ϵ. This will enable us to express the boundary condition at $R = 1$ rather than $R = 1 + \epsilon\cos(\theta)$. This procedure is shown in the following example.

Example 2: *Using a Taylor expansion, find the zeroth-order and first-order boundary conditions.*

Solution: *First, we expand the boundary condition in a Taylor series. Recall the formula for a Taylor series*

$$U(R+h,\theta) = U(R) + h\frac{\partial U}{\partial R}(R,\theta) + \frac{h^2}{2!}\frac{\partial^2 U}{\partial R^2}(R,\theta) + O(h^3).$$

Putting $R = 1$ and $h = \epsilon\cos(\theta)$ and going only to terms of order ϵ^1, we obtain

$$U(1,\theta) + \epsilon\cos(\theta)\frac{\partial U}{\partial R}(1,\theta) + \ldots = 0. \tag{8}$$

Now we can substitute the form of the perturbation series (4) into each term of (8) to obtain

$$U_0(1,\theta) + \epsilon U_1(1,\theta) + \ldots$$
$$+ \epsilon\cos(\theta)\left(\frac{\partial U_0}{\partial R}(1,\theta) + \epsilon\frac{\partial U_1}{\partial R}(1,\theta) + \ldots\right) + \ldots = 1.$$

We now collect terms of each power of ϵ. For ϵ^0 we obtain

$$U_0(1,\theta) = 1 \tag{9}$$

and, for ϵ^1 we obtain

$$U_1(1,\theta) = -\cos(\theta)\frac{\partial U_0}{\partial R}(1,\theta). \tag{10}$$

Solving the zeroth-order equations

From (6) and (9) we have the governing equations for the zeroth-order term U_0,

$$\frac{1}{R}\frac{\partial}{\partial R}\left(R\frac{\partial U_0}{\partial R}\right) + \frac{1}{R^2}\frac{\partial^2 U_0}{\partial\theta^2} = -1, \qquad U_0(1,\theta) = 1.$$

We also have an implicit boundary condition that U_0 is finite at $R = 0$.

The symmetry of this problem suggests that U_0 is independent of θ. The problem then reduces to solving a simple ordinary differential equation for U_0.

Example 3: *Assuming a form of the solution $U_0(R, \theta) = f(R)$, obtain the zeroth-order solution.*

Solution: *Substituting $U(R, \theta) = F(R)$ into (6) gives*

$$\frac{1}{R}\frac{d}{dR}\left(R\frac{dF}{dR}\right) = -1.$$

We can obtain the general solution here by integrating twice. Multiplying through by R and integrating once gives

$$R\frac{dF}{dR} = \frac{R^2}{2} + c_1$$

where c_1 is an arbitrary constant. Dividing through by R and then integrating gives

$$f(R) = -\frac{R^2}{4} + c_1 \log R + c_2$$

where c_2 is also an arbitrary constant. Hence

$$U_0(R, \theta) = -\frac{R^2}{4} + c_1 \log R + c_2.$$

We now apply the boundary conditions.

We apply, first, the implicit boundary condition at $R = 0$, which states that the temperature is finite at $R = 0$. We thus see that $c_1 = 0$ since $\log(R) \to -\infty$ as $R \to 0$. Second, applying the boundary condition $U_0(1, \theta) = 1$ gives $-(1/4) + c_2 = 1$, hence $c_2 = 5/4$.

The solution for $U_0(R, \theta)$ is thus

$$U_0(R, \theta) = -\frac{R^2}{4} + \frac{5}{4}. \tag{11}$$

Solving the first-order equations

The first-order equations are the partial differential equation (7) and boundary condition (10). We substitute the zeroth-order solution (11) into the boundary condition (10). This gives the governing equations

$$\frac{1}{R}\frac{\partial}{\partial R}\left(R\frac{\partial U_1}{\partial R}\right) + \frac{1}{R^2}\frac{\partial^2 U_1}{\partial \theta^2} = 0, \qquad U_1(1, \theta) = \frac{1}{2}\cos(\theta). \tag{12}$$

We also have the implicit condition that U_1 is finite at $R = 0$.

The form of the boundary conditions suggests we seek a solution of the form $U_1(R, \theta) = F(R) \cos\theta$, where F is some function of only one variable. Substituting this into the partial differential equation yields the ordinary differential equation for $F(R)$. The details are worked through in the following example.

Example 4: *Solve the PDE and boundary condition (12) by assuming the form of the solution to be $U_1(R, \theta) = F(R) \cos(\theta)$.*

Solution: *Substituting $U_1(R, \theta) = f(R) \cos(\theta)$ into the partial differential equation (12) gives the ordinary differential equation for $f(R)$,*

$$\frac{1}{R}\frac{d}{dR}\left(R\frac{dF}{dR}\right) - \frac{1}{R^2}F = 0.$$

We cannot integrate this so we expand the derivative, using the product rule, to obtain

$$R^2\frac{d^2F}{dR^2} + R\frac{dF}{dR} - F = 0.$$

This is an Euler–Cauchy equation.

  We now look for a solution of the form $F(R) = R^\lambda$. Substituting this into the differential equation yields $\lambda = \pm 1$. A general solution is therefore

$$F(R) = c_3 R + c_4 R^{-1}$$

where c_3 and c_4 are arbitrary constants. Since $U_1(R, \theta) = F(R) \cos(\theta)$ then the general solution for $U_1(R, \theta)$ is

$$U_1(R, \theta) = \left(c_3 R + \frac{c_4}{R}\right)\cos(\theta).$$

  Applying the boundary condition that U_1 is finite at $R = 0$ means that $c_4 = 0$ identically. Applying $U_1(1, \theta) = (1/2)\cos(\theta)$ yields $c_3 = 1/2$. Hence, we obtain

$$U_1(R, \theta) = \frac{1}{2}R\cos(\theta). \tag{13}$$

Discussion

Putting together the zeroth-order solution and the first-order correction we obtain

$$U(R, \theta) = -\frac{R^2}{4} + \frac{5}{4} + \epsilon\frac{1}{2}R\cos(\theta). \tag{14}$$

We note that the small perturbation in the shape of the surface does not change the temperature of the centre $(R = 0)$. In Figure 4.3.1 the temperature on the surface $R = 1 + \epsilon \cos(\theta)$ is plotted, as a function of θ.

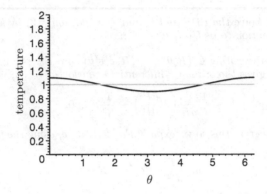

Fig. 4.3.2. Approximate temperature of the centre of a slightly elliptical cylinder given by equation (14). The grey line corresponds to $\epsilon = 0$ and the dark line to $\epsilon = 0.2$.

4.4 Solving the laser drilling case study

We apply the method of perturbations to find an approximate solution to the case study problem of finding the drilling speed of a laser through a thick sheet of metal.

The governing equations

In Section 4.1 we obtained the governing equations and introduced a scaling so that they could be expressed in dimensionless form. From Section 4.1, the scaled governing equations were: the scaled heat conduction equation

$$\frac{\partial U}{\partial T} = \frac{\partial^2 U}{\partial X^2} \qquad (1)$$

with the initial and boundary conditions

$$U(X,0) = 0, \quad U(S(T),T) = 1, \qquad U(\infty,T) = 0 \qquad (2)$$

and the moving boundary condition (modified Stefan condition)

$$\frac{dS}{dT} = 1 + \epsilon\frac{\partial U}{\partial X}(S(T),T). \qquad (3)$$

The dimensionless parameter ϵ is given by

$$\epsilon = \frac{ku_v}{\rho a\lambda} = \frac{cu_v}{\lambda}. \qquad (4)$$

We found $\epsilon \ll 1$ for common metals to be in the range 0.19—0.25. These relatively small (but not too small) values suggest that heat conduction is not the main factor governing the drilling speed, but its presence may still have an effect on the speed.

Perturbation scheme

Let us assume that each of the dependent variables in the problem, U and S, can be expanded in a series of terms involving powers of ϵ. Thus we assume

$$U = U_0 + \epsilon U_1 + \epsilon^2 U_2 + \dots , \qquad (5)$$
$$S = S_0 + \epsilon S_1 + \epsilon^2 S_2 + \dots , \qquad (6)$$

where the functions U_0, U_1, S_0, S_1, etc., are still to be determined.

We substitute (5) and (6) into the governing equations (1-3) and keep terms to $O(\epsilon)$. One difficulty here is that we have to apply boundary conditions on $X = S$, but $S(T)$ is unknown. Since we have assumed that $S(T)$ is a zeroth-order term plus corrections we can use Taylor series to expand the boundary conditions about the zeroth-order solution for the moving boundary S_0. The details of doing this are carried out in the exercises — it follows the method introduced in Section 4.3. We are left with

$$U_0 + \epsilon U_1 + \epsilon S_1\frac{\partial U_0}{\partial X} + \dots = 1 \qquad \text{on } X = S_0(T) \qquad (7)$$

and

$$\frac{dS_0}{dT} + \epsilon\frac{dS_1}{dT} = 1 + \epsilon\frac{\partial U_0}{\partial X} \qquad \text{on } X = S_0(T). \tag{8}$$

Zeroth-order equations

Collecting the terms of order ϵ^0 (see exercises, Question 15) produces the equations for the zeroth-order terms, U_0 and S_0,

$$\frac{\partial U_0}{\partial T} = \frac{\partial^2 U_0}{\partial X^2} \tag{9}$$

with initial and boundary condition

$$U_0(X, 0) = 0, \quad U_0(\infty, T) = 0. \tag{10}$$

We also have

$$U_0 = 1 \quad \text{on } X = S_0, \tag{11}$$

and

$$\frac{dS_0}{dT} = 1. \tag{12}$$

Zeroth-order quasi-equilibrium solution

From equation (12) we can easily integrate to obtain $S_0 = T + c_1$ where c_1 is an arbitrary constant. Since the moving boundary starts from $x = 0$ at $t = 0$ then $s(0) = 0$ and so $S_0(0) = 0$. Consequently $c_1 = 0$ and hence

$$S_0 = T.$$

Thus we know **explicitly** the position of the moving boundary to zeroth-order. The equations we now have to solve, for the zeroth-order temperature, are the partial differential equation (9) with the initial and boundary conditions (10) and (11), with $S_0 = T$.

This is still a moving boundary problem, but we know the position of the moving boundary at all times. Thus we change the reference frame so that the origin is fixed on the known moving boundary thus turning the problem into a fixed boundary problem. Let us define the new coordinates ξ and τ by

$$\xi = X - T, \qquad \tau = T. \tag{13}$$

Using the chain rule (see exercises, Question 16) the partial differential equation (9) is transformed to

$$\frac{\partial U_0}{\partial \tau} - \frac{\partial U_0}{\partial \xi} = \frac{\partial^2 U_0}{\partial \xi^2}. \tag{14}$$

The initial and boundary conditions (10) and (11) are transformed to

$$U_0(\xi, 0) = 0, \tag{15}$$

$$U_0(0, \tau) = 1, \tag{16}$$

$$U_0(\infty, \tau) \to 0. \tag{17}$$

This set of equations can be solved using the method of Laplace transforms†. However, some insight can be gained by examining large time behaviour. Thus we look for the "quasi-equilibrium" solution, obtained by setting the derivative with respect to τ to zero. This corresponds to the large time solution in the moving coordinate system (ξ, τ). Setting $\partial U_0/\partial \tau = 0$ in (14) gives a simple constant-coefficient differential equation to solve. The solution (see exercises, Question 16), which also satisfies the boundary conditions (16) and (17) is

$$U_0(\xi) = e^{-\xi} \tag{18}$$

Reverting to our original coordinates (X, T), using (13), the zeroth-order solution can be written as

$$U_0 = e^{-(X-T)}.$$

This quasi-equilibrium solution has neglected transients relative to the moving boundary (to zeroth-order).

The solution predicts that the dimensionless temperature dies off very quickly a short distance from the moving boundary. This is expected since we are supplying a large amount of energy to the moving boundary

† It is also possible to obtain the time dependent solution using the method of Laplace transforms, see Andrews and McLone (1976) and Bedding (1994). This solution is

$$U_0(\xi, \tau) = \frac{1}{2} e^{-\xi} \left[1 - \mathrm{erf}\left(\frac{\xi - \tau}{\sqrt{4\tau}}\right) \right] + \frac{1}{2} \left[1 - \mathrm{erf}\left(\frac{\xi + \tau}{\sqrt{4\tau}}\right) \right].$$

Note that the limit $\tau \to \infty$ corresponds to the quasi-equilibrium solution.

from the laser but, in the zeroth-order approximation, we are neglecting conduction of heat away from the moving boundary. The quasi-equilibrium temperature is shown in Figure 4.4.1, for some different times. The temperature shows an exponential decay from the moving boundary into the material.

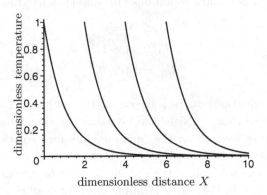

Fig. 4.4.1. Graph showing the zeroth-order, quasi-equilibrium temperature as a function of dimensionless distance X for various values of dimensionless time T; $T = 1$, $T = 2$, $T = 3$ and $T = 4$.

For further interpretation, one should use the fully time-dependent solution, obtained by Laplace transforms, from Andrews and McLone (1976).

First-order correction for the moving boundary

Now let us consider the equations for the first-order correction, which incorporates the effect of conduction. These are obtained from the perturbation expansion of equations (2a–d) (see exercises, Question 15) and collecting terms of order ϵ^1. The first-order terms from the Stefan condition yield

$$\frac{dS_1}{dT} = \frac{\partial U_0}{\partial X}(S_0(T), T).$$

Using the solution for U_0 just obtained, and using $S_0 = T$ this reduces to

$$\frac{dS_1}{dt} = -1.$$

Hence

$$S_1 = -T + c_2$$

where c_2 is an arbitrary constant. But $S_1(0) = 0$ here so

$$S_1 = -T.$$

Thus, since $S = S_0 + \epsilon S_1$, then

$$S(T) = T - \epsilon T.$$

Further insight is obtained by expressing this approximate solution in terms of the original dimensional variables (see Section 4.1). Restoring original variables, using

$$S = \frac{v_0}{\alpha}s, \qquad T = \frac{v_0^2}{\alpha}t,$$

we have

$$s(t) = (1 - \epsilon)v_0 t.$$

In original variables (see Section 4.1) the position of the moving boundary is

$$s(t) = \left(1 - \frac{cu_v}{\lambda}\right)v_0 t = \left(1 - \frac{cu_v}{\lambda}\right)\frac{Q_0}{\rho\lambda A}t.$$

Our new estimate for the drilling speed is

$$v_1 = \frac{ds}{dt} = \left(1 - \frac{cu_v}{\lambda}\right)v_0 = \left(1 - \frac{cu_v}{\lambda}\right)\frac{Q_0}{\rho\lambda A}. \qquad (19)$$

Discussion

We note that both the zeroth-order term and the first-order correction lead to a moving boundary which is proportional to time, rather than to the square root of time, as was the case for the moving boundary problems in Chapter 2. The effect of an increase in the latent heat means that the moving boundary moves more slowly — i.e. the hole is drilled more slowly. This is what we would expect physically. The slower speed comes from a small amount of the power of the laser being conducted away from the moving boundary rather than going towards vaporising the metal. We can also reduce the speed of the moving boundary by increasing the conductivity.

Actually, the speed is not exactly constant — this result was obtained

by using the quasi-equilibrium zeroth-order temperature. A slow variation in time is obtained by using the fully time dependent zeroth-order
solution obtained using Laplace transforms.

Further reading

This case study was sourced from Andrews and McLone (1976). They
also go into more depth in the problem, coming up with a more accurate solution for small times (using the method of matched asymptotic
expansions for a singular perturbation). See also Bedding (1994) for details of the solution using Laplace transforms of the fully time dependent
problem.

Good references for the method of perturbations (including singular perturbations) are Logan (1987), Holmes (1995) and Nayfeh (1981).
Also Aziz and Na (1984) consider perturbations for a number of examples in heat conduction.

4.5 Problems for Chapter 4

1.   *A simple extension of the no-conduction model is to include a term for
the heat lost due to raising the temperature of the metal.*

(a) *Modify the no-conduction model of section 4.1 and hence obtain the
drilling speed estimate v_1 as given by*

$$v_1 = \frac{Q_0}{\rho A(\lambda + cu_v)}.$$

(b) *Given $cu_v = 2.49 \times 10^6 \, \mathrm{J\,kg^{-1}}$ compare the two estimates v_0 and v_1,
where v_0 is the estimate for the drilling speed obtained in Section 4.1.
Use the same values for Q_0, ρ and λ as used in Section 4.1.*

(c) *From the formulae for v_1 and v_0 verify that v_1 is always smaller than
v_0. Give a physical reason for this.*

2.   *Substitute the dimensionless variables*

$$U = \frac{u}{u_v}, \quad T = \frac{v_0^2}{\alpha}t, \quad X = \frac{v_0}{\alpha}x \quad \text{and} \quad S = \frac{v_0}{\alpha}s,$$

into the equations

$$\frac{\partial u}{\partial t} = \alpha \frac{\partial^2 u}{\partial x^2},$$
$$u(x, 0) = 0,$$
$$u(s(t), t) = u_v,$$
$$\rho \lambda \frac{ds}{dt} = \frac{Q}{A} + k \frac{\partial u}{\partial x}(s(t), t),$$

and hence express them in dimensionless form.

3.   *Consider the initial value problem*

$$\frac{dy}{dx} = 1 + (1 + \epsilon)y^2, \qquad y(0) = 0$$

where ϵ is a small parameter. Substitute

$$y(x) = y_0(x) + \epsilon y_1(x) + \cdots$$

*into the differential equation and boundary conditions and collect terms of like order. Solve the resulting **linear** differential equations and hence obtain the approximate solution*

$$y(x) = \tan(x) + \frac{1}{2}\epsilon(x \sec^2(x) - \tan(x)) + \cdots .$$

4.   *The dimensionless, steady-state concentration $C(X)$ satisfies the differential equation*

$$\frac{d^2 C}{dX^2} = \epsilon C^2 + C,$$

with $C(0) = 1$ and $C(\infty) = 0$.

(a) *Use the method of perturbations to find a solution to $O(\epsilon)$.*

(b) *Suppose the boundary condition at $X = 0$ is replaced by $C(0) = 1 - \epsilon$. Will this change C_0 or C_1?*

5.   *For each of the following, use the method of perturbations to find an approximate solution, correct to order ϵ:*

(a) $\dfrac{du}{dt} = 1 - \epsilon u^2$, $u(0) = 2$.

(b) $\dfrac{du}{dt} = u - \epsilon u^2$, $u(0) = 2$.

6.   Consider the differential equation

$$\frac{dx}{dt} = 2\epsilon(x^2 - x) + 1, \qquad x(0) = 0.$$

Given $\epsilon \ll 1$, use the method of perturbations to find the zeroth-order, first-order and second-order terms.

7.   Consider the differential equation

$$\frac{d^2U}{dx^2} - \epsilon U^4, \qquad U(0) = 1, \quad U(1) = 0.$$

For $\epsilon \ll 1$ and assuming $U = U_1(x) + \epsilon U_1(x) + \epsilon^2 U_2(x) + \ldots$ find the second-order term $U_2(x)$. (*The zeroth-order and first-order terms have already been found in Section 4.2.*)

8.   Consider the initial value problem

$$\frac{dy}{dt} + 2y = \epsilon y^2, \qquad y(0) = 3.$$

Solve the initial value problem, for $\epsilon = 0$, to obtain

$$y(t) = 3e^{-2t}.$$

Using the method of perturbations, setting $y = y_0 + \epsilon y_1$, find the first-order correction, $y_1(t)$, for the initial value problem in (a), with $\epsilon \neq 0$ and $\epsilon \ll 1$.

9.   The equation for a nonlinear oscillator (the Duffing equation) is given by

$$\frac{d^2x}{dt^2} + x - \epsilon x^3 = 0,$$

with the initial conditions $x(0) = 1$, $\dot{x}(0) = 0$.

(a) Find the zeroth-order perturbation solution.

(b) Find the solution for the $O(\epsilon)$ term.

(c) Write down (but do not solve) the differential equations and initial conditions for the $O(\epsilon^2)$ term.

(d) Given the solution for x_1 in (b) is

$$x_1(t) = -\frac{1}{32}\cos(3t) + \frac{1}{32}\cos(t) + \frac{3}{8}t\sin(t),$$

comment on the validity of the approximation for large t.

10. *Attempt to apply the method of perturbations (to $O(\epsilon)$) to the differential equation*

$$\epsilon \frac{d^2U}{dX^2} + (1+\epsilon)\frac{dU}{dX} + U = 0, \quad U(0) = 0, \quad U(1) = 1,$$

*where ϵ is a small parameter. What goes wrong? Find the **exact** solution and sketch this for $\epsilon = 0.01$. Comment on the size of the term $\epsilon u''$ as X gets smaller.*

[Note: This type of problem is called a singular perturbation problem. Special methods involving matching inner and outer expansions must be employed to properly handle such problems.]

11. *In the problems for Chapter 1, Question 20, the scaled equations for a missile in an inverse-square gravitational field were*

$$\ddot{X} = \frac{-1}{(1+\epsilon X)^2}, \quad X(0) = 0, \quad \dot{X}(0) = 1$$

where ϵ is a small parameter. Find, correct to $O(\epsilon)$, an approximate solution using the method of perturbations.

12. *Consider the partial differential equation*

$$\frac{1}{r}\frac{\partial}{\partial r}\left(r\frac{\partial u}{\partial r}\right) = -2$$

with boundary condition

$$u(1+\epsilon, \theta) = 10.$$

(a) *Assuming the solution is also finite at $r = 0$ use the method of perturbations to find the zeroth-order and first-order approximate solutions.*

(b) *Give a physical interpretation for the PDE and boundary conditions.*

13. *In fluid mechanics, plane water waves satisfy Laplace's equation $\nabla^2\phi = 0$ for the velocity potential ϕ where the horizontal and vertical velocity components are given by $v_1 = \partial\phi/\partial x$ and $v_2 = \partial\phi/\partial y$. On the surface of the wave $y = h(x,t)$ (see Figure 4.5.1) the boundary conditions are*

$$\frac{\partial\phi}{\partial t} + v_1\frac{\partial\phi}{\partial x} = v_2 \quad \text{and} \quad \frac{\partial\phi}{\partial t} + \frac{1}{2}v^2 = gh.$$

The first equation is the kinematic condition and the second comes from applying Bernoulli's equation.

In dimensionless variables X, T, H and Φ, corresponding to x, t, h and ϕ, these equations may be expressed as $\nabla^2\Phi = 0$ with the boundary conditions

$$\frac{\partial H}{\partial T} + \epsilon\frac{\partial\Phi}{\partial X}\frac{\partial H}{\partial X} = \frac{\partial\Phi}{\partial Y} \quad \text{on } Y = \epsilon H(X,T)$$

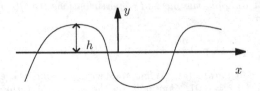

Fig. 4.5.1. Diagram for Question 13. Coordinate system for a water wave. The mean position is located at $y = 0$.

and

$$\frac{\partial \Phi}{\partial T} + \frac{\epsilon}{2}\left(\left(\frac{\partial \Phi}{\partial X}\right)^2 + \left(\frac{\partial \Phi}{\partial Y}\right)^2\right) = \beta H, \quad \text{on } Y = \epsilon H(X, T)$$

*where $\epsilon = h_0/\lambda$ is the (small) ratio of the amplitude to the wavelength and β is a dimensionless constant called the **Froude number**.*

  (a) Assume perturbation expansions $H = H_0 + \epsilon H_1 + \ldots$ and $\Phi = \Phi_0 + \epsilon \Phi_1 + \ldots$ and deduce that the zeroth-order terms satisfy

$$\frac{\partial H_0}{\partial T} = \frac{\partial \Phi_0}{\partial Y}, \quad \frac{\partial \Phi_0}{\partial T} = \beta H_0 \quad \text{on } Y = 0.$$

[Note: In dimensional variables these equations correspond to

$$\frac{\partial h}{\partial t} = \frac{\partial \phi}{\partial y} \quad \text{and} \quad \frac{\partial \phi}{\partial t} = gh \quad \text{on} \quad y = 0$$

which are the well-known equations for linear wave theory.]

14.   *(Continuation of Question 13.) Write down (but do not solve) the equations for the the first-order perturbation for the wave height H_1. [These equations give a means for studying nonlinear effects on wave shape.]*

15.   *In the case study problem of Section 4.4 the dimensionless equations are*

$$\frac{\partial U}{\partial T} = \frac{\partial^2 U}{\partial X^2},$$

with

$$U(S(T), T) = 1$$

and

$$\frac{dS}{dT} = 1 + \epsilon \frac{\partial U}{\partial X}(S(T), T),$$

where ϵ is a small parameter. Make the substitutions

$$U = U_0 + \epsilon U_1 + \epsilon^2 U_2 + \ldots \quad \text{and} \quad S = S_0 + \epsilon S_1 + \epsilon^2 S_2 + \ldots.$$

(a) *Collecting terms, obtain the zeroth-order equations and the first-order equations and solve both of them.*

(b) *Write down the second-order equations (but do not solve).*

16. The zeroth-order perturbation in Question 15 satisfies

$$\frac{\partial U_0}{\partial T} = \frac{\partial^2 U_0}{\partial X^2}$$

with $U_0(X, 0) = 0$, $U_0(T, T) = 1$ and $U_0(\infty, T) = 0$.

(a) *Using a change of variables, $\xi = X - T$, $\tau = T$, show that the governing equation becomes*

$$\frac{\partial U_0}{\partial \tau} - \frac{\partial U_0}{\partial \xi} = \frac{\partial^2 U_0}{\partial \xi^2},$$

and express the initial and boundary conditions in terms of the new variables.

(b) *Obtain a quasi-steady-state solution by neglecting the derivative with respect to τ. Express this back in terms of the variables X and T. Using sketches, explain what this tells us about the long-term behaviour of the temperature.*

5

Case Study: Factory Fires

We consider the spontaneous ignition of sawdust layers in a particle board factory. We introduce bifurcation analysis as a technique for determining the critical value of a parameter (in this case the thickness of the sawdust layer) where the solution suddenly jumps to a different branch (corresponding to ignition).

5.1 Introduction to the case study problem

We introduce here a case study about fires in a chipboard factory. Our aim is to develop equations which can be used to determine conditions for spontaneous ignition to occur. This leads to critical dimensionless combustion parameters that can be evaluated from real reactivity and thermal transport data. The outcome is a criterion for safe storage of mildly combustible materials. The case study is based on an article by Sisson (1993).

Problem background

A chipboard factory has had several unexplained fires. The question arises whether these may have started from spontaneous ignition of sawdust piling up on top of the hot presses (see Figure 5.1.1).

If the heat cannot diffuse fast enough through the sawdust then dangerously high temperatures may result. The thicker the pile of sawdust,

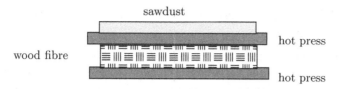

Fig. 5.1.1. Diagram showing a wood press. Fires may be caused by spontaneous ignition of the sawdust piled on top of the hot presses.

the more likely ignition will occur. The essential question is to determine the critical thickness of the sawdust where spontaneous ignition
will occur.

The chemical reaction of oxygen with the sawdust and machinery lu-
bricant produces heat internally. The rate of heat produced will depend
on the chemical reaction and on the temperature — the higher the temperature, the more heat is produced.

Mathematical model

To investigate this problem we try to simplify it. We consider a one-
dimensional heat flow model for a pile of sawdust of height ℓ. This is
shown in Figure 5.1.2.

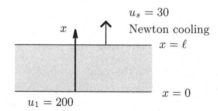

Fig. 5.1.2. One-dimensional heat flow model for pile of sawdust.

The governing equation for this problem is the heat equation with a
volumetric heat source term Q, for the chemical reaction. This equation
is

$$\rho c \frac{\partial u}{\partial t} = k \frac{\partial^2 u}{\partial x^2} + Q \qquad (1)$$

where k is the heat conductivity, ρ the density and c the specific heat of the sawdust.

The boundary $x = 0$ corresponds to where the sawdust sits on the hot metal press. A suitable boundary condition specifies the temperature of the sawdust to be the same as the temperature of the press, which is at $u_1 = 200\,°\mathrm{C}$. We will thus write

$$u(0, t) = u_1. \tag{2}$$

The top surface of the sawdust is exposed to air. We shall assume this surface loses heat to the air according to Newton's law of cooling. (Recall that Newton's law of cooling gives the heat flux as proportional to the temperature difference between the surface and the surroundings.) We thus write

$$-k\frac{\partial u}{\partial x}(\ell, t) = h(u(\ell, t) - u_s) \tag{3}$$

where k is the conductivity of sawdust, h is the Newton cooling coefficient and u_s is the air temperature. We shall investigate several different air temperatures, from $u_a = 20\,°\mathrm{C}$ to $u_a = 40\,°\mathrm{C}$.

The Arrhenius law

We need to determine the form of the heat production term Q in (1). This term represents the rate of heat produced per unit time per unit mass of sawdust. This will be given by the product of the sawdust concentration multiplied by the rate of reaction, K (which is strongly temperature dependent). Heat is released at a rate proportional to the rate of reaction. We have defined the quantity Q as the rate of heat released per unit mass of reactant per unit time. We can write

$$Q = CQ_0K$$

where C is concentration of the reactant, and Q_0 is a positive constant called the heat of reaction.

The **Arrhenius law** in chemistry (after Arrhenius, circa 1900) gives the reaction rate K as a function of temperature. The form of the Arrhenius law is

$$K = Ae^{-E/(Ru)}$$

where u is the temperature, measured in kelvin, E is the activation energy for the reaction, R is the universal gas constant and the constant A is called the pre-exponential factor. Note that A has dimensions $[A] = T^{-1}$, so that $1/A$ is a time scale. The quantity E/R has dimensions of temperature. This is a temperature at which a fraction $1/e$ of reactant particles will react. In practice, this temperature will not be reached before unstable generation of heat results in ignition.

We will assume that the concentration of reactants remains constant over the time interval we are interested in. Thus, if C_0 is the initial concentration of reactant, we obtain

$$Q(u) = C_0 Q_0 A e^{-E/(Ru)}. \tag{4}$$

The Arrhenius law takes the same form as the Boltzmann–Gibbs probability for a particle having sufficient energy to overcome the single reaction energy barrier E, at temperature u.

As we can see in Figure 5.1.3 the rate of heat produced is zero if the temperature is absolute zero — there is no reaction at this temperature. The rate of heat production increases with temperature, rapidly at first. Eventually there is a saturation effect as the rate of heat produced asymptotes to a constant value Q_0 for very large temperatures. There is a point of inflection at $u = 2E/R$.

For many problems in combustion the value of the constant E/R is typically of the order 10^4 K or higher. For sawdust, $E/R \simeq 16,220$ K.

The Frank-Kamenetskii approximation

The $1/u$ term in the Arrhenius law causes some mathematical difficulties due to the difficulty in integrating $e^{-1/u}$. Frank-Kamenetskii's aim was to produce an approximation to the Arrhenius term which is based around a temperature u_1 but still had the exponential characteristic of the original Arrhenius term. The approximation is essentially taking a Taylor-series expansion of the argument of the exponential in the Arrhenius term.

To approximate $e^{-E/(Ru)}$ about some temperature u_1 we can use a

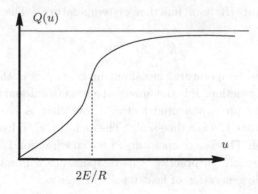

Fig. 5.1.3. Rate of heat produced, $Q(u)$, per unit time per unit mass, in an exothermic chemical reaction according to the Arrhenius law. This is a schematic diagram, not drawn to scale.

two-term Taylor expansion of $1/u$ about $u = u_1$. Recall the Taylor-series formula, for some function f,

$$f(u) = f(u_1) + f'(u_1)(u - u_1) + \ldots + \frac{1}{n!}f^{(n)}(u_1)(u - u_1)^n + \ldots.$$

Hence, for $f(u) = u^{-1}$, and retaining only two terms, we obtain

$$\frac{1}{u} \simeq \frac{1}{u_1} - \frac{1}{u_1^2}(u - u_1).$$

Hence

$$\frac{E}{Ru} \simeq \frac{2E}{Ru_1} - \frac{E}{Ru_1^2}u.$$

Substituting this into the expression for $Q(u)$ we obtain

$$Q(u) \simeq \gamma e^{bu} \tag{5}$$

where

$$b = \frac{E}{Ru_1^2} \quad \text{and} \quad \gamma = C_0 Q_0 e^{-2bu_1}. \tag{6}$$

This is a closely fitting model for the exponentially rising part of the curve in Figure 5.1.3, where $u \ll 2E/R$. As remarked earlier in this section, this is a very high temperature which is not likely to occur before a criterion for ignition has already been established. Hence, for our purposes, it is not important that (5) does not closely approximate the Arrhenius relationship in the region $u \geq 2E/R$.

Summary of governing equations

Using the Frank-Kamenetskii approximation for the Arrhenius reaction rate, the governing partial differential equation for the temperature inside the sawdust is

$$\rho c \frac{\partial u}{\partial t} = k \frac{\partial^2 u}{\partial x^2} + \gamma e^{bu} \tag{7}$$

where the constants b and γ are defined by (6). We also have the two boundary conditions

$$u(0,t) = u_1, \qquad -k\frac{\partial u}{\partial x}(L,t) = h(u(L,t) - u_s). \tag{8}$$

Our aim here is to determine conditions where ignition occurs. We will interpret ignition as a sudden change in temperature as we vary some parameter (e.g. the length ℓ or the air temperature u_a).

Two simpler problems

Before we attempt to solve this we will look at two simpler problems. In each problem we will investigate how the equilibrium temperature changes as we vary some parameter λ. This will give us the necessary skills to study the case study problem defined by the above governing equations.

In the first problem we will look at a problem without any heat conduction. To mimic the physical processes for the case study we will assume that heat is produced by a reaction term and that there is also a loss term, according to Newton's law of cooling. The temperature is now independent of x and we have an ordinary differential equation for the temperature.

In the second problem we will study the simplest heat conduction problem with spontaneous ignition. This problem has one advantage where, by symmetry, we know the maximum temperature occurs at $x = 0$.

5.2 Bifurcations and spontaneous ignition

We investigate the basic phenomenon of spontaneous ignition by studying the mathematics of bifurcation of solutions in an ordinary differential equation. We consider the equilibrium (steady-state) solutions and determine a condition for spontaneous ignition when the equilibrium solutions undergo a sudden jump as some parameter is varied smoothly.

Example problem

A simple example which mimics ignition phenomena is

$$\frac{du}{dt} = -u + \lambda e^u. \tag{1}$$

Here λ is a positive constant, u is absolute temperature and t is time. This is similar to the governing equation obtained in Section 5.1, equation (7), with the heat conduction term missing.

We are interested in how the solution varies as the parameter λ is changed. Physically, changing λ might correspond to changing some physical quantity, such as the amount of the reactant or the ambient air temperature. In the case study problem, the parameter we shall vary is the thickness of the sawdust pile.

As well as being a crude model for the case study problem (neglecting conduction), this differential equation also arises from the study of a **stirred chemical reactor**, see Figure 5.2.1. In a stirred chemical reactor, chemicals are mixed together in a vat and heat is produced by the reaction. Because the mixture is stirred the temperature is homogeneous in space and therefore heat conduction is neglected. However, heat is lost from the surface of the vat to the surrounding air. Spontaneous ignition can occur when the volume of the mixture exceeds a critical value. In the exercises (see Question 1) the derivation of the governing differential equation is worked through and using scaling, equation (1) is obtained.

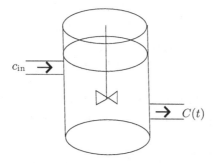

Fig. 5.2.1. A stirred chemical reactor. The stirring means that heat conduction can be neglected. Reactants enter at concentration c_{in}.

Numerical investigation

Let us start by comparing numerical solutions of this equation, for some different values of the parameter λ. This is done using the computer package Maple (using the `dsolve` command). The results are shown in Figure 5.2.2.

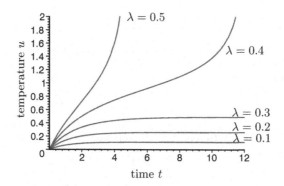

Fig. 5.2.2. Numerical solution of the differential equation (1) for some different values of the parameter λ. Some solutions ($\lambda = 0.1, 0.2, 0.3$) tend to an equilibrium while others ($\lambda = 0.5, 0.5$) do not.

Under some conditions the temperature u will reach an equilibrium, but sometimes this may not occur. It all depends on the value of the parameter λ. In this example, it appears from Figure 5.2.2 that the temperature tends to an equilibrium value for values of λ less or equal than 0.3 but not for values of λ greater than or equal to 0.4. Between 0.3 and 0.4 a critical change in behaviour is occuring. The sudden change in the equilibrium behaviour as a parameter is varied comes from a phenomenon known as a ***bifurcation***. In a bifurcation, it is possible to have more than one equilibrium solution, where the solution can suddenly change from one equilibrium branch to another as the parameter is changed slightly. In the above example, we see a reduction (to zero) of the possible number of equilibrium solutions as λ increases (there is no equilibrium as the temperature tends to infinity).

Equilibrium temperature

In some circumstances the system may approach an equilibrium (steady-state). It means physically that heat is lost from the surface at exactly the same rate as that produced by the chemical reaction. To find the equilibrium temperature we substitute $du/dt = 0$ into the differential equation (1).

Example 1: *Determine graphically the number of equilibrium solutions of (1) as the parameter λ is changed.*

Solution: *Setting $du/dt = 0$ in (1) gives*

$$u = \lambda e^u \tag{2}$$

which can be thought of as an equation for determining the equilibrium values of the temperature u as a function of the parameter λ. First we write the equation in the form

$$\frac{u}{\lambda} = e^u$$

so we can easily sketch both sides of the equation. Equilibrium solutions correspond graphically to intersection points of the two curves

$$y = \frac{u}{\lambda} \quad and \quad y = e^u.$$

The intersection points are shown graphically in Figure 5.2.3.

*The curve $y = e^u$ is monotonic increasing and for large u increases much more rapidly than the curve $y = \lambda^{-1}u$. This proves there will either be **two** solutions (see Figure 5.2.3) for certain values of λ, or **no** solutions for other values (and **one** solution when the two curves intersect with the same tangent).*

Starting with "small" values of λ (large values of λ^{-1}), Figure 5.2.3 demonstrates there are always **two** intersection points. For "small" λ

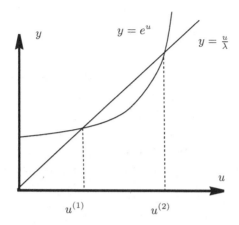

Fig. 5.2.3. Intersection points corresponding to equilibrium temperatures. If the parameter λ is sufficiently large, there are no intersection points.

we can think of these solutions as corresponding to a "cool" equilibrium temperature, which we denote by $u^{(1)}$, and a "hot" equilibrium temperature $u^{(2)}$.

As the parameter λ increases (and the straight line slope λ^{-1} decreases) then we progress to a state where there are no longer any solutions of equation (2) (as shown in Figure 5.2.3).

There is a critical value of the parameter λ where we go from two solutions to no solutions. We shall denote this value by λ_{cr} and the corresponding critical equilibrium temperature by u_{cr}. We see, graphically, that at λ_{cr} the curves intersect and the tangents of the two curves are the same.

Example 2:   *By solving the appropriate equations, determine analytically u_{cr} and λ_{cr}.*

Solution: *From the original equation for the intersection of the two graphs, putting $u = u_{\mathrm{cr}}$ and $\lambda = \lambda_{\mathrm{cr}}$ gives*

$$\frac{u_{\mathrm{cr}}}{\lambda_{\mathrm{cr}}} = e^{u_{\mathrm{cr}}}. \tag{3}$$

Differentiating both sides with respect to u_{cr} gives the equation for the tangents to coincide,

$$\frac{1}{\lambda_{\mathrm{cr}}} = e^{u_{\mathrm{cr}}}. \tag{4}$$

We can solve (3) and (4) simultaneously, to obtain $u_{cr} = 1$. Substituting this
back into one of the equations yields

$$u_{cr} = 1, \qquad \lambda_{cr} = e^{-1} \simeq 0.36788. \tag{5}$$

This gives the exact analytic values for the critical values of λ and
u. In more complicated examples we would probably have to solve the
equations graphically or numerically. The critical value λ_{cr} corresponds
to when spontaneous ignition occurs. But to see why, we need to first
understand which of the two equilibrium solutions the temperature tends
to, in different circumstances. This is known as determining the **stability** of the equilibrium temperatures.

Stability of equilibria

We now determine which of these equilibrium solutions the temperature
tends towards. We can do this by examining the sign of the du/dt term
in the differential equation (1). This indicates what happens to the
temperature for a single given value of λ. The following example shows
how to determine the stability of the equilibrium solutions illustrated in
Figure 5.2.3.

Example 3: *Suppose the value of λ is such that there are two equilibrium
solutions. Determine, graphically, which of them are stable and which are
unstable. Assume the initial temperature is such that it is below both equilibrium temperatures.*

Solution: *We are somewhere in region I in Figure 5.2.4. Everywhere in this
region, $du/dt > 0$ because heat production is always greater than heat loss.
This implies that the temperature must increase in this region. It increases
towards the "cool" equilibrium temperature.*

*Now let us suppose that the initial temperature is such that u is in region
II in Figure 5.2.4. Here heat loss is greater than heat production so the
temperature will decrease towards the "cool" steady-state.*

*Now suppose that the initial temperature u is greater than the "hot" steady-
state (corresponding to region III in Figure 5.2.4). Here heat production is
always greater than heat loss so that the temperature increases without bound.*

A **stable** equilibrium point u_e is one for which the solution $u(t)$ can
be made to remain as close to u_e as we please by taking the initial value

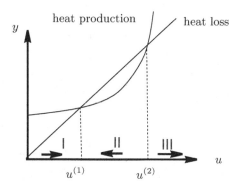

Fig. 5.2.4. For equation (1) with "high" values of λ. There is one "cool" equilibrium temperature and one "hot" equilibrium temperature. However, the temperature always tends towards the "cool" equilibrium temperature, as explained in the text.

$u(0)$ sufficiently close to u_e. An equilibrium point u_e that is not stable is said to be **unstable.**

A stable equilibrium point is **asymptotically stable** if every solution $u(t)$ approaches u_e as $t \to \infty$ provided the initial value $u(0)$ is sufficiently close to u_e. Since the lower-temperature equilibrium point is approached from either side as t increases, it is asymptotically stable. The higher-temperature equilibrium point is unstable because solutions with initial temperature higher than $u^{(2)}$ show a further increase beyond $u^{(2)}$.

Bifurcation diagrams

For a fixed value of λ, such that λ is large, the temperature tends here to the "cool" equilibrium for all initial temperatures greater than 0 and less than $u^{(2)}$. When λ is in the range such that there are no steady solutions for u we easily deduce that $du/dT > 0$. Thus, for these values of λ the temperature increases without bound.

Let us now plot the equilibrium solutions as the parameter λ increases. This is called a **bifurcation diagram.** This is done in Figure 5.2.5. A bifurcation diagram is a convenient way of summarising the information about how the equilibrium solutions change as a parameter changes. In the bifurcation diagram in Figure 5.2.5 ignition occurs where the two solutions become one solution and then zero.

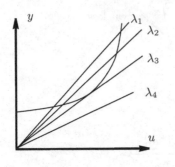
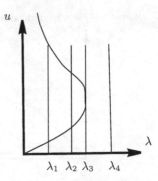

Fig. 5.2.5. Constructing a bifurcation diagram. The bifurcation diagram (b) is a graph of the equilibrium solutions plotted as a function of the parameter λ.

Imagine that the value of λ is such that we are just about to jump from two steady-state solutions to no steady-state solution. A sudden change in the number of steady-state solutions as we vary a parameter slightly is called a ***bifurcation***. As λ is increased slightly past λ_{cr}, we have established that the temperature can suddenly go from approaching a "cool" steady-state to where the temperature increases indefinitely. This sudden rise is interpreted physically as the mixture undergoing spontaneous ignition when the volume of the tank is increased past some critical value.

Further discussion

Spontaneous ignition occurs when the equilibrium temperature suddenly changes from a cool steady-state to a much hotter one (or to where the temperature increases without bound). This can happen if the initial temperature is sufficiently high. However, ignition can, in some cases, also occur spontaneously even for ***all*** initial temperatures.

In real life, the temperature cannot actually become infinite. This is an artefact of the model, caused by using the Frank-Kamenetskii approximation of the heat produced by the reaction. Recall that the Arrhenius term has a saturation effect as the temperature becomes larger, i.e. $Q(u) = C_0 Q_0 e^{E/(Ru)}$. In the exercises (see Question 10), an example with a full Arrhenius term will be investigated. In the example, the temperature jumps from one stable "cool" equilibrium to a much hotter stable equilibrium, as some parameter λ changes slowly.

In general, a change in the number, or stability, of equilibrium solutions is said to be a **bifurcation**. This change may be effected when some system parameter (for example, λ) reaches some critical value. The system parameters are usually expressed as dimensionless quantities so that critical values of the system parameters are independent of the system of units used. A bifurcation diagram is often useful for identifying where bifurcations occur, and for determining whether they correspond physically to spontaneous ignition.

5.3 Ignition with conduction

In the previous section we looked at a problem where conduction of heat was not important because the temperature was homogeneous in space. A problem where conduction is important is solved here as a precursor to the case study problem.

Example problem

Let us consider a reacting material between two surfaces which are both maintained at temperature u_1. We place the origin at the centre of the reacting material with the two surfaces at $x = \pm 1$ (see Figure 5.3.1). As the material between the two surfaces reacts with oxygen, heat is released, which causes the reaction to accelerate. However, heat is also conducted through the material and escapes to the surroundings. With more material between the surfaces, it will be more difficult for heat to escape by conduction through the material.

Due to the symmetry of the problem, the maximum temperature will occur at the origin. To find conditions under which spontaneous ignition can occur we develop equations for the maximum temperature and use these to construct a bifurcation diagram which shows how the maximum temperature inside the material varies as the distance between the surfaces changes.

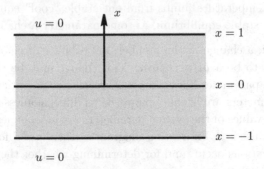

Fig. 5.3.1. Combustible material between two surfaces $x = \pm 1$, both held at temperature $u = 0$.

Governing equations

The governing equation for this problem is then the heat equation with a source term given by the Arrhenius law with the Frank-Kamenetskii approximation. This equation is

$$\rho c \frac{\partial u}{\partial t} = k \frac{\partial^2 u}{\partial x^2} + \gamma e^{bu}$$

where γ and b are positive constants. Here k is the heat conductivity, ρ the density and c the specific heat of the material.

For equilibrium temperature $\partial u / \partial t = 0$ and u is then a function of x alone. Hence an equilibrium temperature $u(x)$ satisfies

$$k \frac{d^2 u}{dx^2} + \gamma e^{bu} = 0. \tag{1}$$

We also assume that the boundaries of the medium are at $x = \pm \ell$ and at these boundaries the temperature is kept constant at some value u_1. However, the symmetry of this problem allows us to work with only half the region.

For mathematical simplicity, we let $u_1 = 0$, $b = 1$, $\ell = 1$, and let $\lambda = \gamma / k$, which gives the differential equation

$$\frac{d^2 u}{dx^2} + \lambda e^u = 0, \tag{2}$$

with the boundary conditions

$$\frac{du}{dx}(0) = 0, \quad u(1) = 0. \tag{3}$$

We show in the exercises (see Question 12) that this is equivalent to a certain scaling to make the equations dimensionless.

Our aim is to determine a condition on λ for spontaneous ignition to occur. For this we will examine how the *maximum* equilibrium temperature changes with the parameter λ.

For this problem, by symmetry, the maximum temperature must occur at $x = 0$, where we have

$$u(0) = u_m. \tag{4}$$

Solving the differential equation

The differential equation (2) is nonlinear. However, because the differential equation doesn't depend on the x-variable explicitly this suggests we proceed by eliminating x using the chain rule. The following example shows how to obtain the general solution.

Example 1: *Use the chain rule to solve the differential equation (2).*

Solution: *Letting $u' = du/dx$, and using the chain rule,*

$$\frac{d^2u}{dx^2} = \frac{du'}{dx} = \frac{du'}{du}\frac{du}{dx} = u'\frac{du'}{du},$$

the differential equation (2) becomes

$$u'\frac{du'}{du} = -\lambda e^u.$$

Solving this first-order separable differential equation we obtain

$$\frac{1}{2}(u')^2 = -\lambda e^u + C_1$$

where C_1 is an arbitrary constant of integration. Taking the square root of both sides, we obtain

$$\frac{1}{\sqrt{2}}\frac{du}{dx} = \pm\sqrt{C_1 - \lambda e^u}.$$

We discard the positive sign for the following reason. The highest temperature will be in the middle at $x = 0$ so the temperature must decrease with increasing x. This implies that $du/dx < 0$ and so we can discard the positive case. Thus we have

$$\frac{1}{\sqrt{2}}\frac{du}{dx} = -\sqrt{C_1 - \lambda e^u}.$$

We can now apply the remaining boundary conditions and then obtain $u(x)$ by solving the remaining first-order differential equation. The following example shows how to do this.

Example 2: *Apply the boundary conditions and complete the solution of the differential equation.*

Solution: *Applying the condition (3), we can evaluate the arbitrary constant C_1. Substituting $u = u_m$ gives $C_1 = \lambda e^{u_m}$. We thus obtain*

$$\frac{du}{dx} = -\sqrt{(2\lambda(e^{u_m} - e^u))}.$$

This is a first-order separable differential equation.

The solution of this differential equation may be written as

$$\int \frac{1}{\sqrt{e^{u_m} - e^u}}\, du = -\sqrt{2\lambda}\,x + C_2 \tag{5}$$

where C_2 is an arbitrary constant of integration. The integral can be evaluated in closed form (see Question 13 in the exercises). We obtain†

$$\frac{-2}{e^{u_m/2}}\operatorname{sech}^{-1}\left(\frac{e^{u/2}}{e^{u_m/2}}\right) = -\sqrt{2\lambda}\,x + C_2. \tag{6}$$

With some algebraic manipulation, and applying the boundary condition $u(0) = u_m$, we can solve for u to get

$$u(x) = u_m - 2\log\left[\cosh\left(\sqrt{\frac{\lambda}{2}}e^{u_m/2}x\right)\right]. \tag{7}$$

† Note that $\operatorname{sech}(z) = 1/\cosh(z)$. The function sech^{-1} is the inverse of the sech function. It may also be denoted arcsech.

We still have one undetermined constant in the solution, u_m, which is the temperature at the centre $x = 0$. To determine u_m we can apply the remaining boundary condition (3), $u(1) = 0$. This gives the equation

$$u_m - 2 \log \left(\cosh \left(\sqrt{\frac{\lambda}{2}} e^{u_m/2} \right) \right) = 0. \tag{8}$$

We can, in principle, solve this equation for u_m, where we think of u_m as depending on the parameter λ.

Ignition

We can now solve equation (8), either graphically or numerically, to determine u_m as a function of the parameter λ. The software package Maple is a useful tool for this task. In Figure 5.3.2 a graph of the LHS of this equation is given, for some different values of λ. For the lower values of λ there appear to be two solutions. For the larger value there is no longer any solution. A bifurcation has occurred. We can see the critical value of λ occurs at just over 0.8. This behaviour mimics that of the simple example in the previous section.

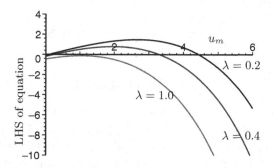

Fig. 5.3.2. Graph of the LHS of equation (8) versus u_m for some different values of λ. The solutions of equation (8) occur where the curve crosses the horizontal axis.

In this simple example, we can carry out some algebraic manipulation of (8) to put the unknown u_m on one side of the equation,

$$e^{-u_m/2} \cosh^{-1}(e^{u_m/2}) = \sqrt{\frac{\lambda}{2}} \tag{9}$$

This form allows us to prove there are at most two equilibrium solutions. A graph of the LHS is given in Figure 5.3.3.

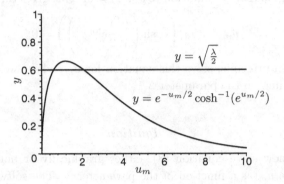

Fig. 5.3.3. Graphs of both sides of equation (9). The intersection points are the solutions, corresponding to equilibrium temperatures.

It is also possible to obtain the critical value of λ more accurately. The critical value of λ occurs where the two intersection points with $y = \text{constant}$ suddenly become one, i.e. where the curve in Figure 5.3.3 is a maximum, giving

$$\sqrt{\frac{\lambda_{\text{cr}}}{2}} \simeq 0.663, \qquad \Rightarrow \lambda_{\text{cr}} \simeq 0.88. \tag{10}$$

Bifurcation diagram

Using Maple we can also construct the bifurcation diagram by plotting the two solutions for u_m as we vary the parameter λ. This is done with the `implicitplot` command. This result is shown in Figure 5.3.4. From this we can also read off the critical value of λ as approximately 0.88.

An examination of the stability of the two equilibrium solutions is beyond the scope of this book — it is more difficult for partial differential equations. However, it has been shown (see, for example, Weber and Renkema (1995)) that the upper branch is unstable whereas the lower branch is stable. This is as we might expect from the physically similar example in the previous section in which the conduction term of the current model was replaced by a simpler loss term.

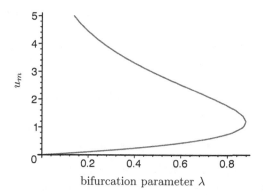

Fig. 5.3.4. Bifurcation diagram. This was obtained using the Maple function `implicitplot`.

Extensions

This problem, in scaled variables, is instructive, but needs to be generalised to where we can apply more complicated boundary conditions, such as for the case study problem. We also need to consider what to do when the temperatures of the top and the bottom are not the same. When they are the same, then symmetry allows us to assume that the maximum temperature occurs at $x = 0$.

To handle these more complicated problems, one can introduce a new variable $x_{\max}$ denoting the position where the maximum temperature occurs. Then, solving the problem as before, we use one boundary condition to determine the critical λ and use the other to find the value of $x_{\max}$. This is what we have to do in the next section.

Also, it is possible to solve similar problems to (2) in cylindrical and spherical geometries,

$$\frac{1}{r}\frac{d}{dr}\left(r\frac{du}{dr}\right) + \lambda e^u = 0, \qquad u(1) = 0, \tag{11}$$

$$\frac{1}{r^2}\frac{d}{dr}\left(r^2\frac{du}{dr}\right) + \lambda e^u = 0, \qquad u(1) = 0. \tag{12}$$

For these it is possible to obtain the critical values

$$\lambda_{\mathrm{cr}} = 2.0, \qquad \lambda_{\mathrm{cr}} \simeq 3.32$$

respectively, see Jones (1993).

5.4 Solving the factory fire case study

In this section we provide a solution for the case study problem. This solution follows a similar approach to that developed in the previous two sections, that of creating a bifurcation diagram, with spontaneous combustion occurring when the sawdust layer is too large.

Review of case study problem

Sawdust piling up on the surfaces of hot presses is thought to be the cause of fires in a chipboard factory. In Section 5.1 we developed the equations for a one-dimensional model for the temperature in a pile of sawdust. The model is shown in Figure 5.4.1.

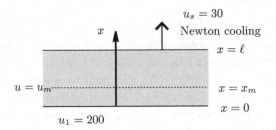

Fig. 5.4.1. One-dimensional model for the temperature in a pile of sawdust.

The temperature $u(x,t)$ satisfies the heat equation with a volumetric heat source due to the oxidation reaction with the sawdust. With the Frank-Kamenetskii approximation of the Arrhenius expression for the rate of heat production we obtained the partial differential equation

$$\rho c \frac{\partial u}{\partial t} = k \frac{\partial^2 u}{\partial x^2} + \gamma e^{bu} \qquad (1)$$

where b and γ are positive constants. These constants are given by (see Section 5.1)

$$\gamma = C_0 Q_0 e^{-2bu_1} \qquad b = \frac{E}{R u_1^2}. \qquad (2)$$

For the equilibrium temperature we set $\partial u/\partial t = 0$ to obtain

$$k\frac{d^2u}{dx^2} + \gamma e^{bu} = 0. \tag{3}$$

On the metal surface $x = 0$ we have a prescribed temperature $u_1 = 200\,°C$ and we apply Newton cooling on the top surface $x = \ell_2$. Thus we write

$$u(0) = u_1, \qquad -k\frac{du}{dx}(\ell) = h(u(\ell) - u_s) \tag{4}$$

where $u_s = 30\,°C$ is the temperature of the surroundings, k is the conductivity and h the Newton cooling coefficient.

Solution for temperature

In the problem studied in the previous section we knew the maximum temperature occurred at the line of symmetry. In this problem we do not know, in advance, where the maximum temperature occurs. Let us assume that it occurs at the point $x = x_m$, to be determined. Defining the maximum temperature as u_m we have

$$\frac{du}{dx} = 0 \qquad \text{when} \quad u = u_m \quad \text{at} \quad x = x_m. \tag{5}$$

Solving (3) for the equilibrium temperature follows the same procedure as in Section 5.3. The details are left to the exercises. The solution, satisfying (5) is

$$u(x) = u_m - \frac{2}{b}\log\left[\cosh\left(\sigma e^{bu_m/2}(x - x_m)\right)\right] \tag{6}$$

where the positive constant σ is defined by

$$\sigma = \sqrt{\frac{b\gamma}{2k}}. \tag{7}$$

This solution has two unknown constants, u_m and x_m, which need to be determined by applying the boundary conditions at $x = 0$ and $x = \ell$.

Applying the boundary conditions

Using the boundary condition (4), $u = u_1$ at $x = 0$, on (6) gives the equation

$$u_1 = u_m - \frac{2}{b}\log\left[\cosh\left(\sigma e^{bu_m/2}x_m\right)\right] \qquad (8)$$

since cosh is an even function. This gives an equation for u_m but we still need an additional equation for x_m.

This second equation comes from applying the Newton cooling condition on $x = \ell$, given in (4). This gives (see exercises, Question 17) the equation

$$\frac{k}{h}e^{bu_m/2}\sigma\tanh\left(\sigma e^{bu_m/2}(\ell - x_m)\right) =$$
$$\frac{b}{2}(u_m - u_s) - \log\left[\cosh\left(\sigma e^{bu_m/2}(\ell - x_m)\right)\right]. \qquad (9)$$

This now gives us two equations for the two unknowns u_m and x_m both of which depend on the parameter ℓ.

Bifurcation diagram

We need to substitute values for the constants b and σ, defined by equations (2) and (7). Appropriate data for sawdust is:

$$\frac{E}{R} = 16,220\,\text{K}, \qquad k = 0.103\,\text{W m}^{-1}\text{K}^{-1},$$

$$C_0Q_0 = 7.1 \times 10^{17}\,\text{J m}^{-3}, \qquad \frac{h}{k} = 80\,\text{m}^{-1}.$$

With $u_1 = 200\,°\text{C} = 473\,\text{K}$ we thus calculate $b \simeq 0.072\,\text{K}^{-1}$ and $\sigma \simeq 6.4 \times 10^{-7}\,\text{m}^{-1}$.

Equations (8) and (9), solved simultaneously, define u_m and x_m as a function of ℓ. Maple can be used to find possible solutions of the pair of equations (8) and (9), for various values of ℓ. We can now use the Maple function `implicitplot` to sketch the solutions for u_m as ℓ varies. This is the *bifurcation diagram* for u_m. This is shown in Figure 5.4.2. In the diagram on the left there appears to be only one solution for u_m (with a corresponding solution for x_m) with these values of the parameters, but for sufficiently large ℓ there are no solutions. Note that we have only captured one branch of the bifurcation diagram. We need to refine the diagram to see the other branch. This is also shown in Figure 5.4.2

where we have "zoomed in" and plotted the bifurcation diagram over a smaller temperature range.

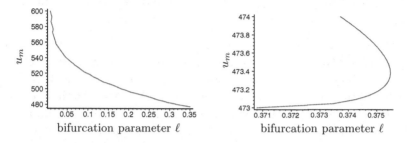

Fig. 5.4.2. Two views of the bifurcation diagram obtained using the Maple function `implicitplot` (i.e. solution of equations (8) and (9) for u_m and x_m, at various values of ℓ). In the diagram on the left we have searched for temperature values $473\,\mathrm{K} = 200\,^\circ\mathrm{C}$ to $573\,\mathrm{K} = 300\,^\circ\mathrm{C}$ which showed only the upper branch of the bifurcation diagram. In the diagram on the right a more refined search was made which yielded both branches. The critical value can now be read off as $\ell_{\mathrm{cr}} \simeq 0.378$ from the diagram on the right as the value of ℓ where the curve begins to turn back towards the vertical axis.

The critical value of ℓ, where the equilibrium solutions cease to exist, occurs at approximately

$$\ell_{\mathrm{cr}} \simeq 0.38\,\mathrm{m} = 38\,\mathrm{cm}.$$

This suggests that, for the values of the parameters used here, ignition will occur if the dust layer exceeds 38 cm.

Alternatively, the bifurcation diagram can also be obtained by first solving (8) explicitly for x_m, yielding

$$x_m = \sigma^{-1} e^{-bu_m/2} \cosh^{-1}\left(e^{b(u_m - u_1)/2}\right). \tag{10}$$

Substituting the result back into (9) gives an equation of the general form

$$F(u_m, \ell) = 0$$

which now defines u_m implicitly as a function of ℓ. Again the Maple function `implicitplot` can be used to obtain the bifurcation diagram.

Further reading

This case study was sourced from Sisson (1993). A good introduction to the mathematics of spontaneous ignition is given in Logan (1987) with a more advanced discussion of chemical reactors in Fowler (1997). Jones (1993) gives a broader view of the area of combustion science, as does Drysdale (1985). The original development of the theory of spontaneous ignition with conduction is given in Frank-Kamenetskii (1972).

For a further application of the theory of spontaneous combustion to industry see Gray (1988) which discusses possible spontaneous ignition of dust on cylindrical power cables. An extension of this to banks of cables is treated in Weber and Renkema (1995). Also, McNabb et al. (1999) considers spontaneous combustion in coal pillars. A problem involving microwave heating, with some mathematical similarities to the spontaneous combustion equations is discussed in Hill and Smyth (1990).

5.5 Problems for Chapter 5

1.   *In a stirred chemical reactor the reactants are continually stirred together so that the temperature within is uniform. There is heat loss from the surface of the container, given by Newton's law of cooling (see Figure 5.2.1).*

(a) *Deduce the differential equation*

$$V c \rho \frac{du}{dt} = V Q(u) - hA(u - u_a)$$

from a suitable "word equation" expressing conservation of heat. Here the temperature of the mixture is u, V is the volume of the mixture and c is the specific heat, $Q(u)$ denotes the rate of heat produced per unit volume per unit time and u_a is the temperature of the surroundings.

(b) *Let $Q(u) = \gamma e^{bu}$, corresponding to a Frank-Kamenetskii approximation of an Arrhenius heat production term. Using the scaling*

$$U = b(u - u_a), \qquad T = \frac{hA}{\rho c V} t$$

deduce that the scaled temperature U satisfies

$$\frac{dU}{dT} = \lambda e^{U} - U,$$

where λ is a positive constant (give λ in terms of the other physical constants).

2. An exothermic reaction takes place in a tank of volume V and is fed by a stream of constant flow rate q (volume per unit time) of constant reactant concentration c_{in} and constant temperature u_{in}. The mixture is removed from the tank with the same flow velocity q (see Figure 5.2.1).

Assume that the reactant is used up at the rate $KQ_0Ce^{-B/u}$ where $C(t)$ is the concentration of the reactant and $u(t)$ is the temperature, and where $B = E/R$ and Q_0 and K are positive constants. The heat, per unit volume of mixture, released is $CQ_0e^{-B/u}$.

(a) By carrying out both a heat and mass balance, derive ordinary differential equations for C and u.

(b) Introduce the dimensionless concentration z and dimensionless temperature θ and dimensionless time T defined by

$$T = \frac{t}{V/q}, \qquad z = \frac{C}{c_{in}}, \qquad \theta = \frac{u}{u_{in}}$$

and hence show that the equations obtained in (a) become

$$\frac{dz}{dT} = 1 - z - \lambda z e^{-\gamma/\theta}, \qquad \frac{d\theta}{dT} = 1 - \theta + \lambda bz e^{-\gamma/\theta}.$$

Identify the dimensionless constants λ, μ and b.

3. Consider the two differential equations obtained in Question 2 for the dimensionless temperature θ and dimensionless concentration z in a continuous stirred reactor with consumption of reactant.

(a) By eliminating $e^{-\gamma/\theta}$ from the differential equations and using the initial conditions deduce that

$$\frac{d}{dT}(\theta + bz) = 1 + b - (\theta + bz).$$

(b) Given the initial conditions $\theta(0) = 1$ and $z(0) = 1$ deduce that $\theta + bz = 1 + b$ for all T. Explain where the initial conditions come from.

(c) Hence deduce that θ satisfies

$$\frac{d\theta}{dT} = 1 - \theta + \lambda(1 + b - \theta)e^{-\gamma/\theta} = 0.$$

4. Consider the ordinary differential equation

$$\frac{du}{dt} = \lambda e^{3u} - (u - 1),$$

where λ is a positive constant and $u > 0$.

(a) Deduce that there exist two equilibrium solutions provided the value of λ is sufficiently small, and there are no equilibrium solutions if λ is above some critical value. What is this critical value?

(b) What happens to the temperature if λ is small?

(c) If λ is large, deduce that the temperature u tends to the lower equilibrium temperature [Hint: Consider the stability.]

5.   Consider the ordinary differential equation

$$\frac{du}{dt} = \lambda(u^2 + 1) - u.$$

(a) Deduce that there exist two equilibrium solutions provided the value of λ is sufficiently small, and there are no equilibrium solutions if λ is above some critical value. What is this critical value?

(b) What happens to the temperature if λ is small?

(c) If λ is large, deduce that the temperature u tends to the lower steady-state temperature.

6.   Sketch a bifurcation diagram for equilibrium solutions of

$$\frac{du}{dt} = e^u - \lambda u.$$

Determine which branches are unstable.

7.   Consider the differential equation

$$\frac{du}{dt} = \lambda u^2 - u$$

where λ is a positive constant and $u > 0$.

(a) Determine graphically the number of equilibrium solutions as λ varies from 0 to ∞.

(b) Give a rough sketch of the bifurcation diagram. Determine which branches are unstable.

(c) Obtain, analytically, an expression for λ_{cr}.

8.   Consider the differential equation

$$\frac{du}{dt} = \lambda u - u.$$

Are there any bifurcations of the equilibrium solutions?

9.   Consider

$$\frac{du}{dt} = \lambda e^{-u/(u+1)} - u$$

where λ is a positive constant. If u denotes a dimensionless temperature, determine the critical value of λ such that spontaneous ignition occurs.

10. If we choose not to approximate the Arrhenius term with the Frank-Kamenetskii approximation, the differential equation is (in certain dimensionless variables)

$$\frac{du}{dt} = \lambda e^{-\beta/u} - (u - 1)$$

where β is a positive dimensionless constant and $u > 0$.

(a) Sketch $e^{-\beta/u}$ as a function of u. [Hint: Show there is a point of inflection at $u = \beta/2$.]

(b) Graphically, determine the number of equilibrium solutions as λ increases from 0 to ∞.

(c) Hence, give a rough sketch of the bifurcation diagram. Determine which branches are unstable.

(d) For $\beta = 10$, find the numerical value of λ_{cr} where spontaneous ignition first occurs as we vary λ.

11. Consider the equilibrium expression

$$1 - \theta + \lambda(1 + b - \theta)e^{-\gamma/\theta} = 0$$

from the differential equation for θ derived in Question 3.

(a) Defining the function $f(\theta) = (1 + b - \theta)e^{-\gamma/\theta}$ show that f has a turning point and a point of inflection. Hence sketch the graph on $(0, \infty)$.

(b) Determine the number of equilibrium solutions as the parameter λ ranges from 0 to ∞.

(c) Hence sketch the bifurcation diagram.

(d) The phenomenon of **quenching** is where the temperature of a reaction is decreased suddenly when the flow rate of reactants q is slowly increased past some critical value. Discuss whether this model incorporates quenching. [Note: λ is inversely proportional to q. Refer to Question 2 for background.]

12. Consider the differential equation

$$k\frac{d^2u}{dx^2} + \gamma e^{bu} = 0, \qquad u(\pm\ell) = u_1.$$

Define the dimensionless variables $U = b(u - u_1)$ and $X = x/\ell$. Show the problem may be written in the dimensionless form

$$\frac{d^2U}{dx^2} + \lambda e^U = 0, \qquad U(\pm 1) = 0.$$

Give the dimensionless constant λ in terms of the other physical constants in the problem.

13. From the integral in equation (5), in Section 5.3, show that

$$\int \frac{1}{\sqrt{e^{u_m} - e^u}}\, du = \frac{-2}{e^{u_m/2}}\, \text{sech}^{-1}\left(\frac{e^{u/2}}{e^{u_m/2}}\right).$$

[Hint: Make the substitutions $e^u = v^2$ then $v = a\,\text{sech}(y)$ where $a^2 = e^{u_m}$.]

14. Given the general solution (6) from Section 5.3

$$\frac{-2}{e^{u_m/2}}\, \text{sech}^{-1}\left(\frac{e^{u/2}}{e^{u_m/2}}\right) = -\sqrt{2\lambda}\,x + C_2$$

apply the boundary condition and obtain

$$u = u_m - 2\log\left[\cosh\left(\sqrt{\frac{\lambda}{2}}e^{u_m/2}x\right)\right].$$

15. Solve the differential equation, arising from the case study in Section 4.4,

$$k\frac{d^2 u}{dx^2} + \gamma e^{bu} = 0$$

with the appropriate boundary conditions, to obtain

$$u(x) = u_m - \frac{2}{b}\log\left[\cosh\left(\sigma e^{b u_m/2}(x - x_m)\right)\right], \qquad \sigma = \sqrt{\frac{b\gamma}{2k}}.$$

You may assume the integral

$$\int \frac{1}{\sqrt{a^2 - e^u}}\, du = \frac{-2}{a}\, \text{sech}^{-1}\left(\frac{e^{u/2}}{a}\right).$$

16. In Section 5.3 we obtained the equation

$$u_m = 2\log\left[\cosh\left(\sqrt{\frac{\lambda}{2}}e^{u_m/2}\right)\right].$$

(a) Show that this can be written as

$$e^{-u_m/2}\cosh^{-1}(e^{u_m/2}) = \sqrt{\frac{\lambda}{2}}.$$

(b) By graphing the LHS find the critical value of λ.

17. Consider

$$\frac{\partial u}{\partial t} = \frac{\partial^2 u}{\partial x^2} + \lambda e^u$$

with the boundary conditions

$$u(0, t) = 0, \qquad u(1, t) = 10.$$

(a) Assuming the maximum temperature u_m occurs at $x = x_{\max}$, obtain two simultaneous equations for $x_{\max}$ and u_m, both of which depend on the parameter λ.

(b) By plotting, find the approximate critical value of λ where spontaneous ignition occurs.

6

Case Study: Irrigation

We consider a problem from the agricultural industry which involves calculating the optimal irrigation furrow width to enable the soil moisture content to be sufficiently high where a crop is planted. We introduce two mathematical techniques needed to solve this problem — the first uses the Kirchoff transformation to transform a nonlinear PDE into a linear one; the second uses a Fourier-series expansion which takes advantage of the periodic nature of the problem to solve the linear PDE.

6.1 Introduction to the case study problem

We introduce the case study problem of designing optimal spacing of irrigation furrows. We also develop the relevant equations for the water content in the soil. This case study is based on consulting experiences of one of the authors.

Background

In many countries irrigation systems are used to raise crops in areas with low rainfall. To irrigate rows of crops, shallow furrows are placed between the crops along which water is supplied.

In this case study, we will set up a model whose aim is to predict optimal spacing of irrigation furrows. We can control the water flow-rate and would like this to be as small as possible. The problem is to determine the maximum distance between the furrows and the width of the furrows such that there is sufficient water available for the crops.

172

Fig. 6.1.1. Syphon irrigation of onion crops in the Murrumbidgee Irrigation Area near Griffith, NSW, Australia. Photography by Bill van Aken, © CSIRO Land and Water.

This requires water to be available at sufficient concentration and also requires that the water is not so tightly bound into the pores that it cannot be extracted by the plant roots.

In dry soils a pressure must be applied to extract the water from small pores. This pressure, the suction pressure, is caused by capillary forces. A tensiometer is a device used to measure the suction pressure. It is a tube filled with water with a semi-permeable cap which passes water but not air. This is inserted into the soil and it causes a drop in the water level due to the suction pressure. This drop in water level is called the *suction potential* (or sometimes the capillary potential) and we denote it here by the symbol Ψ. The suction potential is related to the suction pressure p by $p = \rho g \Psi$, where ρ is density of water and g is the acceleration due to gravity. Note that Ψ is negative, corresponding to a suction pressure, with $\Psi = 0$ when the soil is saturated. Plants are able to raise water to their full height H through capillary action. Similarly, they are able to extract water that is held in the soil with a suction potential of the order of $-H$. In this case study we consider a crop plant that can extract water from soils provided the suction potential is greater than -150 cm, see, for example Rowell (1994).

The model

We consider an array of equally spaced irrigation furrows, each of width $2w$. We suppose that water is supplied at a rate of R units of volume per unit furrow length per unit time. The furrows are spaced a distance 2ℓ apart. This is shown in Figure 6.1.2 below. We let z denote distance down into the soil and choose $x = \pm\ell$ to be lines of symmetry.

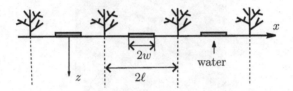

Fig. 6.1.2. A periodic array of rows of crops, spaced a distance 2ℓ apart. At the centre of each row is an irrigation furrow of width $2w$.

We assume that there is a large number of furrows. Near a central furrow, the outer furrows will have negligible influence. Hence, the number of furrows can be assumed to be infinite. We also assume the furrows are very long, effectively infinite in length. Then the flow pattern will be two-dimensional (in the x and z directions).

The moisture content (volume of water to total volume) is denoted by θ which will be a function of x, z and time t. By symmetry, we need only consider the semi-infinite strip defined by $-\ell < x < \ell$ and $0 < z < \infty$. This is shown in Figure 6.1.3.

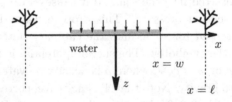

Fig. 6.1.3. Diagram showing coordinate system for model.

Governing equations

Flow in porous media is governed by Darcy's law. Darcy's law states that the water volume flux $\mathbf{v}$ is proportional to the gradient of the pressure,

$$\mathbf{v} = -\frac{K_s}{\rho g}\nabla(p - \rho g z), \tag{1}$$

where ρ is the density of water, g is the acceleration due to gravity and K_s is a constant, for each soil, called the **hydraulic conductivity**. The term $\rho g z$ represents the gravitational contribution to the overall pressure.

For unsaturated flow, the hydraulic conductivity is a function of the moisture content θ. Writing $p = \rho g \Psi$, Darcy's law is extended to

$$\mathbf{v} = -K(\theta)\nabla(\Psi - z). \tag{2}$$

The hydraulic conductivity $K(\theta)$ is a strongly increasing function of the moisture content θ. Typically, $K(\theta)$ can decrease by several orders of magnitude for only a 40% decrease in θ. This reflects how it is much harder to drive water through smaller pores.

The starting point for developing a governing equation for the moisture content in the soil is conservation of mass applied to a small arbitrary volume of width δx, height δz and depth L. The rate of change of mass (or volume, since density is constant) is equated to the net mass flux (or volume flux) into and out of the volume. This yields the equation (see exercises, Question 1)

$$\frac{\partial \theta}{\partial t} = -\left(\frac{\partial v_1}{\partial x} + \frac{\partial v_2}{\partial z}\right). \tag{3}$$

Substituting the modified form of Darcy's law (2) into the conservation of mass equation (3) we obtain

$$\frac{\partial \theta}{\partial t} = \frac{\partial}{\partial x}\left(K(\theta)\frac{\partial \Psi}{\partial x}\right) + \frac{\partial}{\partial z}\left(K(\theta)\frac{\partial \Psi}{\partial z}\right) - \frac{\partial}{\partial z}(K(\theta)). \tag{4}$$

Let us define $D(\theta) = K(\theta)\, d\Psi/d\theta$. Then using the chain rule, we obtain

$$\frac{\partial \theta}{\partial t} = \frac{\partial}{\partial x}\left(D(\theta)\frac{\partial \theta}{\partial x}\right) + \frac{\partial}{\partial z}\left(D(\theta)\frac{\partial \theta}{\partial z}\right) - K'(\theta)\frac{\partial \theta}{\partial z}. \tag{5}$$

This equation is known as ***Richards' equation.*** We shall neglect transients and consider the ***equilibrium*** moisture content $\theta(x, z)$. Richards' equation now becomes

$$\frac{\partial}{\partial x}\left(D(\theta)\frac{\partial \theta}{\partial x}\right) + \frac{\partial}{\partial z}\left(D(\theta)\frac{\partial \theta}{\partial z}\right) - K'(\theta)\frac{\partial \theta}{\partial z} = 0. \qquad (6)$$

This PDE is nonlinear. One of the first approximations a mathematician might think about making, here, is to set $D(\theta)$ and $K'(\theta)$ to be constants, resulting in a linear, constant coefficient PDE. However, $K(\theta)$, and hence $D(\theta)$, varies by several orders of magnitude for a small variation in θ, so this approach is inappropriate. There is another way to obtain a linear, constant coefficient equation from the equilibrium Richards' equation.

Boundary conditions

We need to specify boundary conditions on the boundaries $z = 0$, $z \to \infty$. On $z = 0$ we specify the water input. This is in the form of a given water flux in the irrigated part of the furrow, from $x = 0$ to $x = w$. Defining $v_3(x, z)$ as the vertical water flux, then

$$v_3(x, 0) = \begin{cases} R, & 0 < |x| < w, \\ 0, & w < |x| < \ell, \end{cases} \qquad (7)$$

where R is the infiltration rate. We also expect the water concentration to tend to zero as we go deeper into the soil. Thus

$$\theta(x, \infty) \quad \text{is finite.}$$

We also require the moisture content $\theta(x, z)$ to be periodic in the x direction, with period 2ℓ.

6.2 The Kirchhoff transformation

We introduce the Kirchhoff transformation as a method for transforming the nonlinear Richards' equation, from Section 6.1, into a linear partial differential equation.

Definition of the Kirchhoff variable

In the previous section, we obtained the equilibrium Richards equation for the moisture content in a soil. This equation was

$$\frac{\partial}{\partial x}\left(D(\theta)\frac{\partial\theta}{\partial x}\right) + \frac{\partial}{\partial z}\left(D(\theta)\frac{\partial\theta}{\partial z}\right) - K'(\theta)\frac{\partial\theta}{\partial z} = 0. \tag{1}$$

Here $D(\theta) = K(\theta)\Psi'(\theta)$ where θ is the moisture content, K is the hydraulic conductivity and Ψ is the suction potential (capillary potential).

To take advantage of techniques for solving linear partial differential equations it is desirable to transform (6) into a linear partial differential equation. We introduce the new variable $\mu(x,z)$ defined by

$$\frac{\partial\mu}{\partial x} = D(\theta)\frac{\partial\theta}{\partial x}, \qquad \frac{\partial\mu}{\partial z} = D(\theta)\frac{\partial\theta}{\partial z} \tag{2}$$

This is equivalent to

$$\mu(x,z) = \int_{\theta_0}^{\theta} D(\theta)\, d\theta$$

where θ_0 is some reference value of the moisture content θ.

Transforming Richards' equation

By the chain rule

$$\frac{dK}{d\theta} = \frac{dK}{d\theta}\frac{\partial\theta}{\partial z} = \frac{dK}{d\mu}\frac{\partial\mu}{\partial z}. \tag{3}$$

Hence Richards equation now becomes

$$\frac{\partial^2\mu}{\partial x^2} + \frac{\partial^2\mu}{\partial z^2} - K'(\mu)\frac{\partial\mu}{\partial z} = 0. \tag{4}$$

If we now choose

$$K'(\mu) = \alpha$$

then

$$\frac{\partial^2\mu}{\partial x^2} + \frac{\partial^2\mu}{\partial z^2} - \alpha\frac{\partial\mu}{\partial z} = 0 \tag{5}$$

which is a linear partial differential equation (with constant coefficients). This is much easier to solve!

We can also show (see exercises, Question 3) that the assumption $K'(\mu) = \alpha$ is equivalent to

$$K = K_s e^{\alpha \Psi} = \alpha \mu \tag{6}$$

where K_s is the saturated hydraulic conductivity, with $K = K_s$ when $\Psi = 0$. This is, fortunately, reasonably consistent with experimental measurements of hydraulic conductivity verses suction potential. This equation can be used to convert from the artificial variable back to the suction potential Ψ.

Transformed boundary conditions

We need to specify boundary conditions on the boundaries $z = 0$, $z \to \infty$. On $z = 0$ we specify the water input

$$v_3(x, 0) = \begin{cases} R, & 0 < |x| < w, \\ 0, & w < |x| < \ell, \end{cases} \tag{7}$$

where R is the infiltration rate.

We can also write the boundary conditions in terms of the Kirchhoff variable $\mu(x, z)$. Using the definition of μ, (2), Darcy's law (Section 6.1, equation 1) and (6), the boundary condition (7) now becomes

$$-\frac{\partial \mu}{\partial z}(x, 0) + \alpha \mu(x, 0) = \begin{cases} R, & 0 < |x| < w, \\ 0, & w < |x| < \ell. \end{cases} \tag{8}$$

We also require $\mu(x, \infty)$ to be finite and $\mu(x, z)$ to be periodic in x with period ℓ.

6.3 Fourier series solutions

Now we see how to use a Fourier series to solve a linear partial differential equation. The problem chosen to illustrate the method is a simplified version of the linear partial differential equation from Section 6.2.

Example problem

Before solving the case study problem in the next section, we consider a simpler problem. Consider the partial differential equation

$$\frac{\partial^2 \mu}{\partial x^2} + \frac{\partial^2 \mu}{\partial z^2} = 0 \tag{1}$$

with the boundary conditions

$$\mu(x, 0) = \begin{cases} 0, & 0 < |x| < 1, \\ 3, & 1 < |x| < 2. \end{cases} \tag{2}$$

We also assume $\mu(x, z)$ remains finite as $z \to \infty$. We similarly assume the function is periodic in the x-direction, with period 4.

The solution involves assuming a suitable form of $\mu(x, z)$. We shall assume a form which accounts for the periodic variation in the x-variable. For this we use a Fourier series.

Fourier series

Fourier series are used to represent functions that are periodic on some interval. The functions $\sin(ax)$ and $\cos(ax)$ both have period $2\pi/a$, hence $\sin(\pi x/L)$ and $\cos(\pi x/L)$ have period $2L$. Also, the functions $\sin(n\pi x/L)$ and $\cos(n\pi x/L)$, where n is an integer, both have period $2L$ (as well as period $2L/n$). In fact we can represent any function of period $2L$ by an infinite linear combination of these functions (see, for example, Mei (1997) for further details).

The general Fourier-series expansion of a function f on an interval $[-L, L]$, and with period $2L$, is defined by

$$f(x) = \frac{a_0}{2} + \sum_{n=1}^{\infty} a_n \cos\left(\frac{n\pi x}{L}\right) + b_n \sin\left(\frac{n\pi x}{L}\right). \tag{3}$$

The coefficients a_n and b_n are given by†

$$a_n = \frac{1}{L} \int_{-L}^{L} f(x) \cos\left(\frac{n\pi x}{L}\right) dx, \quad b_n = \frac{1}{L} \int_{-L}^{L} f(x) \sin\left(\frac{n\pi x}{L}\right) dx. \tag{4}$$

---

† More generally, a periodic function defined on the interval $[c, c+2L]$ has coefficients given by

$$a_n = \frac{1}{L} \int_{c}^{c+2L} f(x) \cos\left(\frac{n\pi x}{L}\right) dx, \quad b_n = \frac{1}{L} \int_{c}^{c+2L} f(x) \sin\left(\frac{n\pi x}{L}\right) dx.$$

The case $c = -L$ gives the same formulae as above.

The formulae for the b_n Fourier coefficients are easily obtained by multiplying (3) through by $\sin(m\pi x/L)$ and then integrating over the interval $[-L, L]$. Note that, when integrated over this interval, all the terms in the sum become zero except when $m = n$. Similarly the a_n coefficients, including a_0, are obtained by integrating over the same interval, after multiplying through by $\cos(m\pi x/L)$ (see exercises, Question 5).

Solution of example problem

We assume at each depth z the variable μ can be expanded as a Fourier cosine-series where the coefficients are functions of z. Thus, we write

$$\mu(x, z) = \frac{A_0(z)}{2} + \sum_{n=1}^{\infty} A_n(z) \cos\left(\frac{n\pi x}{2}\right) + B_n \sin\left(\frac{n\pi x}{2}\right) \qquad (5)$$

where $A_0(z)$ and $A_n(z)$ are functions of z to be determined. The following example shows how to use this assumed form to solve the partial differential equation.

Example 1: *Solve the PDE (1) by substituting the variable-coefficient Fourier series (5) into the PDE.*

Solution: *The functional form for the coefficients, $A_0(z)$ and $A_n(z)$, is obtained by requiring $\mu(x, z)$ to satisfy the partial differential equation. We substitute this back into the PDE. If we now equate all the coefficients $\sin(n\pi x)$ and $\cos(n\pi x)$ we obtain equations for the coefficients $A_0(z)$ and $A_n(z)$. These are*

$$A_n'' - \frac{n^2 \pi^2}{4} A_n = 0, \qquad B_n'' - \frac{n^2 \pi^2}{4} B_n = 0, \qquad A_0'' = 0.$$

Solving these differential equations gives

$$A_0(z) = a_0 + c_0 z, \qquad (6)$$

$$A_n(z) = a_n e^{-n\pi z/2} + c_n e^{n\pi z/2}, \qquad (7)$$

$$B_n(z) = b_n e^{-n\pi z/2} + d_n e^{n\pi z/2}, \qquad (8)$$

where $a_0, c_0, a_n, b_n, c_n, d_n$ are arbitrary constants.

We also impose the condition that μ be finite as $z \to \infty$. This means that we must choose the constants $b_0 = 0$, $c_n = 0$ and $d_n = 0$. Hence we have

$$A_0(z) = a_0, \qquad (9)$$

$$A_n(z) = a_n e^{-n\pi z/2}, \qquad (10)$$

$$B_n(z) = b_n e^{-n\pi z/2}. \qquad (11)$$

Substituting for $A_0(z)$ and $A_n(z)$ back into (5) we obtain

$$\mu(x, z) = \frac{a_0}{2} + \sum_{n=1}^{\infty} a_n \cos\left(\frac{n\pi x}{2}\right) e^{-n\pi z/2}. \qquad (12)$$

We could also have obtained this solution by looking for separable solutions of the form $\mu(x, z) = X(x)Z(z)$. Substitution into the PDE and application of the separable (homogeneous) boundary conditions yields an infinite number of solutions. The general separated solution (12) is then obtained as a linear combination of these separated solutions (see exercises, Question 4).

Applying the boundary condition (2), and substituting $z = 0$, yields the equation

$$\frac{a_0}{2} + \sum_{n=1}^{\infty} a_n \cos\left(\frac{n\pi x}{2}\right) + b_n \sin\left(\frac{n\pi x}{2}\right) = g(x),$$

where

$$g(x) = \begin{cases} 0, & 0 < |x| < 1, \\ 3, & 1 < |x| < 2. \end{cases}$$

This is equivalent to finding the Fourier series for the RHS.

Example 2: *Find the Fourier series for*

$$g(x) = \begin{cases} 0, & 0 < |x| < 1, \\ 3, & 1 < |x| < 2. \end{cases}$$

Solution: *The function has period $2L = 4$. Hence equation (4) gives*

$$a_n = \frac{1}{2} \int_{-2}^{2} g(x) \cos(n\pi x/2) \, dx.$$

Splitting the integral from 0 to 1 and from 1 to 2, and also using symmetry, we obtain

$$a_n = \frac{2}{2} \int_{1}^{2} 3 \times \sin(n\pi x/2) \, dx = \frac{-6}{n\pi} (\sin(n\pi/2))$$

since $\sin(n\pi) = 0$ for all integer values of n. Similarly we can show $a_0 = 3$ and $b_n = 0$.

We can now substitute the expressions for the coefficients a_0 and a_n back into (12). This gives the solution to the PDE (1) and the boundary condition (2),

$$\mu(x,z) = \frac{3}{2} + \sum_{n=1}^{\infty} \frac{-6\sin(n\pi/2)}{n\pi} \cos\left(\frac{n\pi x}{2}\right) e^{-n\pi z/2}. \qquad (13)$$

6.4 Solving the crop irrigation case study

We solve the case study problem of determining the optimal spacing of irrigation furrows. Some of the details are carried out as exercises in the problems at the end of the chapter.

Review of problem

Rows of crops are irrigated by shallow channels. The water seeps through the soil and moves towards the drier soil where crops are planted. We wish to predict values of moisture content $\theta(x,z)$ and suction potential $\Psi(x,z)$ in the soil, see Figure 6.1.3.

Recall from Section 6.1 that the governing partial differential equation for the moisture content was the equilibrium Richards' equation. In Section 6.2 we saw how to use the Kirchoff transformation to transform the partial differential equation into the constant-coefficient equation

$$\frac{\partial^2 \mu}{\partial x^2} + \frac{\partial^2 \mu}{\partial z^2} - \alpha \frac{\partial \mu}{\partial z} = 0. \qquad (1)$$

We also obtained

$$K = \alpha\mu = K_s e^{\alpha\Psi}, \qquad v_3 = -\frac{\partial \mu}{\partial z} + \alpha\mu. \qquad (2)$$

The boundary conditions were

$$v_3 = -\frac{\partial \mu}{\partial z}(x,0) + \alpha\mu(x,0) = \begin{cases} R, & 0 < |x| < w, \\ 0, & w < |x| < \ell, \end{cases} \qquad (3)$$

and

$$\mu(x,\infty) = 0. \qquad (4)$$

We also require $\mu(x,z)$ to be periodic in x, with period 2ℓ.

Overview of solution

In Section 6.3 we looked at solving a partial differential equation similar to (1). We use the same approach here. We start by assuming a Fourier-series form of the solution

$$\mu(x, z) = \frac{A_0(z)}{2} + \sum_{n=1}^{\infty} \left(A_n(z) \cos \left(\frac{n\pi x}{\ell} \right) + B_n(z) \sin \left(\frac{n\pi x}{\ell} \right) \right) \quad (5)$$

where the functions $A_n(z)$, $B_n(z)$ are to be determined. This gives a periodic solution in the x-direction. Furthermore, symmetry of $\theta(x, z)$ about $x = 0$ also implies symmetry of $\mu(x, z)$ about $x = 0$. This requires $B_n(z) = 0$. Hence

$$\mu(x, z) = \frac{A_0(z)}{2} + \sum_{n=1}^{\infty} A_n(z) \cos \left(\frac{n\pi x}{\ell} \right). \quad (6)$$

We substitute (6) into the PDE (1) to obtain equations for the Fourier coefficients $A_n(z)$, $B_n(z)$. The details are given in the exercises. We then obtain

$$\mu(x, z) = \frac{a_0}{2} + \sum_{n=1}^{\infty} a_n \cos \left(\frac{n\pi x}{\ell} \right) e^{\beta_n z} \quad (7)$$

where a_n and b_n are constants, and

$$\beta_n = \frac{1}{2} \left(\alpha - \sqrt{\alpha^2 + 4n^2\pi^2/\ell^2} \right). \quad (8)$$

Note that all the β_n are negative for all values of $n = 1, 2, \ldots$.

The constants a_n are determined by applying the remaining boundary condition at $z = 0$, given by equation (3). Using the Fourier-coefficients formulae (equation (4) in the previous section) we obtain (see Question 8 in the exercises)

$$a_n = \frac{2R \sin \left(\frac{n\pi w}{\ell} \right)}{n\pi(\alpha - \beta_n)}, \qquad a_0 = \frac{2Rw}{\alpha\ell}.$$

Substituting these back into (7), we obtain the solution for $\mu(x, z)$,

$$\mu(x, z) = \frac{Rw}{\alpha\ell} + 2R \sum_{n=1}^{\infty} \frac{\sin \left(\frac{n\pi w}{\ell} \right)}{n\pi(\alpha - \beta_n)} \cos \left(\frac{n\pi x}{\ell} \right) e^{\beta_n z}, \quad (9)$$

where the constants β_n are defined above. This solution was first obtained by Batu (1978).

Interpretation of the solution

Typical data for loam is

$$\alpha = 0.05 \,\text{cm}^{-1}.$$

(This comes from $\alpha^{-1} = 20\,\text{cm}$ which is a typical scale for the capillary rise in a soil.)

We also assume the spacing between rows of crops as 4 metres apart, so

$$\ell = 200 \,\text{cm}.$$

The vertical flux of water into the soil, R, can be easily measured. We shall take a value of the vertical flux R as

$$R = K_s = 5 \,\text{cm/day}.$$

This quantity is also called the infiltration rate.

In practice, it is found that many crop plants cannot extract water from the soil whenever the suction potential Ψ is more negative than $-150\,\text{cm}$. Let us assume a seed is planted at a depth of 3 cm. Then we require the suction potential to be greater than $-150\,\text{cm}$ at $z = 3\,\text{cm}$. Using this value, we will now use the solution (9) to calculate the minimum value w (the irrigation furrow width) so that the seedling can get enough water to grow. For a seedling planted directly between two irrigation furrows we require

$$\Psi(\ell, 3) > -150 \,\text{cm}.$$

From equation (2), $\alpha\mu = K_s e^{\alpha\Psi}$, which enables us to calculate the suction potential Ψ, given μ, as

$$\Psi(x, z) = \frac{1}{\alpha} \log\left(\frac{\alpha\mu(x, z)}{K_s}\right).$$

Typically, for a clay loam,

$$R = K_s = 5 \,\text{cm/day}.$$

Using the above values, we plot in Figure 6.4.1 values of the suction potential at a depth of 3 cm into the soil, $\Psi(\ell, 3)$, for different values of the irrigation furrow width w.

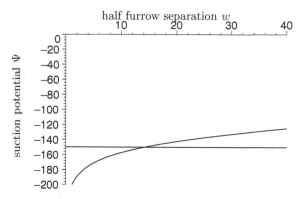

Fig. 6.4.1. Graph of the suction potential Ψ, at depth 3 cm, verses the half width w of the irrigation furrow, with infiltration rate of $R = K_s = 5$ cm/day. The horizontal line is the critical suction potential that seedling roots are capable of overcoming.

Discussion

Referring to Figure 6.4.1, we note that $-\Psi$ increases as w increases. This simply means the suction potential becomes less negative as we make the irrigation furrows wider (and therefore increase the amount of water available) and the plants find it easier to extract water. From Figure 6.4.1 we see that we can maintain a sufficient suction potential of greater than -150 cm at depth 3 cm provided the irrigation furrow width is greater than approximately 28 cm (i.e. $w \simeq 14$ from Figure 6.4.1).

The above analysis is appropriate for seedlings with a low water consumption. For well developed crops, a sink term, representing plant root water extraction, would need to be appended to the RHS of Richards' equation (4) of Section 6.1. The consequences of this are yet to be fully investigated.

Further reading

For a practical discussion of irrigation the reader is referred to Hillel (1982).

A classic reference for hydrology is Bear (1972). Fowler (1997) gives brief treatments of different aspects of porous media flow, concentrating on saturated flows, in particularly the dam seepage problem and pollution leakage. Noble (1967) also contains an elementary discussion of the dam seepage problem. Good references for the mathematics of unsaturated flow in soils are Philip (1970) and Philip (1988).

For a recent discussion of the effect of plant roots on the water flow in unsaturated soils, see Chang and Corapcioglu (1997). For an extension to flow in soils with layers of different hydraulic conductivities, see Stormont and Morris (1997). An application of unsaturated flow in the mining industry is the drainage of coal and iron-ore stockpiles which is discussed in McElwain and Fulford (1995).

6.5 Problems for Chapter 6

1. *Suppose water is flowing in a soil where the velocity of the water at a point (x, z) is $\mathbf{v}(x, z)$, where we have assumed symmetry in the y-direction. Consider the liquid flowing through an arbitrary small box with sides of width δx and δy and depth L. Assume constant density of water ρ.*

(a) *Argue the rate of mass of liquid flowing into the bottom of the box $z = $ constant at z is $\rho v_3(x, z)\delta x L$. Hence determine the total net flux out of the box by considering the flow into or out of the other three sides.*

(b) *Denoting the volume content of the box as $\theta(x, z, t)$ (ratio of volume of water to total volume) determine the rate of change of mass of the liquid.*

(c) *By letting δx and δz both tend to zero hence deduce the equation*

$$\frac{\partial \theta}{\partial t} = -\left(\frac{\partial v_1}{\partial x} + \frac{\partial v_3}{\partial z}\right).$$

2. *Consider the nonlinear diffusion equation*

$$\frac{\partial C}{\partial t} = \frac{\partial}{\partial x}\left(D(C)\frac{\partial C}{\partial x}\right)$$

where the diffusivity $D(C)$ depends on the concentration $C(x, t)$. Define the change of dependent variable

$$h(C) = \int_0^C D(C)\, dC.$$

Show that the nonlinear diffusion equation above is equivalent to

$$\frac{\partial h}{\partial t} = D(C(h))\frac{\partial^2 h}{\partial x^2}.$$

3. Assuming Darcy's law, and the Kirchhoff transformation, show that the assumption $K'(\mu) = \alpha$, where α is a constant, is equivalent to

$$\alpha\mu = K_s e^{\alpha\Psi}.$$

[Hint: First show $\mu = \displaystyle\int_{\Psi_0}^{\Psi} K(\Psi)\,d\Psi$ using the definition of $D(\theta)$.]

4. Consider the partial differential equation

$$\frac{\partial^2\mu}{\partial x^2} + \frac{\partial^2\mu}{\partial z^2} = 0$$

with the boundary conditions

$$\frac{\partial\mu}{\partial x}(0, z) = 0, \qquad \frac{\partial\mu}{\partial x}(1, z) = 0.$$

(a) Assuming a solution of the form $\mu(x, z) = X(x)Z(z)$ deduce that

$$X'' = \lambda X, \quad X'(0) = 0, \quad X'(1) = 0, \qquad Z'' = -\lambda Z,$$

where λ is a separation constant.

(b) By solving the equations for X find values of λ which lead to non-zero solutions.

(c) Hence solve for $Z(z)$, assuming that μ remains finite as $z \to \infty$, and obtain

$$\mu(x, z) = c_0 + \sum_{n=1}^{\infty} c_n \cos(n\pi x)e^{-n\pi z}.$$

5. Consider the Fourier series

$$f(x) = \frac{A_0}{2} + \sum_{n=1}^{\infty} A_n \cos\left(\frac{n\pi x}{L}\right) + B_n \sin\left(\frac{n\pi x}{L}\right),$$

where A_n, B_n, $n = 1, 2, \ldots$ and A_0 are constants. By multiplying this expression through by $\cos(m\pi x/L)$ and integrating over $[-L, L]$ deduce that

$$A_n = \frac{1}{L}\int_{-L}^{L} f(x) \cos\left(\frac{n\pi x}{L}\right)\,dx.$$

Similarly, find an expression for B_n.

6. Solve Laplace's equation

$$\frac{\partial^2\mu}{\partial x^2} + \frac{\partial^2\mu}{\partial z^2} = 0$$

on $-1 < x < 1$, $0 < z < \infty$, with periodic boundary conditions on $x = -1$ and $x = 1$ and with $\mu(x, 0) = 1 - x^2$. Assume the solution is finite as $z \to \infty$.

7.   Consider the partial differential equation

$$\frac{\partial^2 \mu}{\partial x^2} + \frac{\partial^2 \mu}{\partial z^2} - \alpha \frac{\partial \mu}{\partial z} = 0.$$

Assuming a form of the solution

$$\mu(x,z) = \frac{A_0(z)}{2} + \sum_{n=1}^{\infty} A_n(z) \cos\left(\frac{n\pi x}{\ell}\right)$$

and assuming μ remains finite as $z \to \infty$ show that

$$\mu(x,z) = \frac{a_0}{2} + \sum_{n=1}^{\infty} a_n \cos\left(\frac{n\pi x}{\ell}\right) e^{\beta_n z}$$

where a_0, a_n, β_n, $n = 1,2,3,\ldots$, are constants to be determined.

8.   Apply the following boundary condition to the solution obtained in Question 7:

$$-\frac{\partial \mu}{\partial z}(x,0) + \alpha \mu(x,0) = g(x),$$

where

$$g(x) = \begin{cases} R, & 0 < |x| < w, \\ 0, & w < |x| < \ell. \end{cases}$$

7

Conclusions

In this final chapter we review the case studies examined in this book. We also explore (very briefly) some other areas of industry in which mathematics has been extensively used and mention some other mathematical techniques which commonly find application in industry.

7.1 Introduction

In the previous chapters we have explored several case studies from industry. All of these case studies have involved problems which use some variant of the diffusion equation. The case studies were deliberately chosen to use similar mathematics and physical backgrounds to make it easier for the reader and to allow the reader to see the links between the various case studies through mathematics.

In Chapter 2 we considered the problem of continuous casting. This problem introduced the mathematics of moving boundary problems and of similarity solutions using the Boltzmann similarity solution. The problem illustrated how a simplified model involving one dimensional heat flow yielded an exact solution in terms of error functions. Using this exact solution we were able to estimate the size of the puddle of molten steel, and showed that it was of the order of the size of the rotating drum, which meant that the process was not feasible.

Next, in Chapter 3, the case study was from the area of water filtration involving a process known as reverse osmosis. This was a diffusion problem with a non-constant advection coefficient. Here we continued with the idea of a similarity solution, and developed a technique (the method of stretching transformations) as a means for constructing similarity transformations which reduce the dimensionality of a partial differential equation. We were able to construct a solution expressed in terms of an integral which was evaluated numerically. However, the rest of the solution was analytic, giving useful insight into the structure of the solution. This is valuable information to have even if we have available a numerical solution to equations with less restrictive assumptions than those applied here.

In Chapter 4 we considered a problem from manufacturing industry involving the use of lasers to drill holes through metal. We introduced the method of perturbations to obtain a correction to a simple solution for the drilling speed. The correction incorporated the effect of conductivity of the material.

In Chapter 5 the case study involved the spontaneous ignition of wood dust, as found in a particle board factory. The aim of the problem was to determine if spontaneous ignition of the wood dust on top of hot metal presses was possible. The problem was described by a simple heat conduction equation. The mathematics introduced to solve this problem was that of bifurcation of equilibrium solutions.

Finally, in Chapter 6 we considered a problem from agriculture: determining the optimal width of irrigation furrows. The major difficulty with this problem, which involves flows in unsaturated soils, is that the governing equation (Richards' equation) is a highly nonlinear diffusion equation. However, we were able to overcome this by applying a transformation which converted Richards' equation into a standard linear diffusion equation. We also made use of the periodic structure of the problem and used Fourier series to obtain a solution which could be expressed as a sum of terms.

We do not wish the reader to get a false impression that heat or diffusion problems are the only types of problems that involve mathematics in industry. Thus the purpose of this chapter is to expose the reader to the greater variety of mathematical ideas that has been used in industry. As such, we do not intend to explore the topics in nearly the same depth as in the previous chapters, nor do we claim the exposition is a complete coverage of all the areas of mathematics that have been applied in industry. We do hope, however, that the reader will get a good idea of the scope and variety of areas of applications in industry and of the potential use of mathematics as an aid to solving the problems of various industries.

7.2 A survey of mathematical techniques

Mathematics as a discipline has been around since the time of the ancient Greeks and before. Therefore there has been a lot of time to accumulate a huge body of knowledge. It is difficult to imagine how any one person could possibly have sufficient working knowledge of all the techniques available to be able to apply them to a given problem arising in industry. However, this should not deter us from attempting to at least have an awareness of as many techniques as possible. Then, if a given problem demands the use of a technique one is unfamiliar with, if it is important enough to solve the problem, then the technique can be learned.

Within the area of solving diffusion problems, and also other problems involving partial differential equations, there are important techniques we have not used in this book, such as *integral transforms* and *Greens functions*. It is also useful to be aware of extensions of techniques that follow on as extensions of the techniques developed here. These include *Sturm–Liouville* methods (these are an extension to Fourier series), *Lie group methods* and *Bäckland transformations* (extensions of stretching transformations) and various extensions of perturbation methods (e.g. *matched asymptotic expansions*), many of which are covered in more advanced mathematical techniques books, such as Logan (1987).

Three related areas of modelling that are worth becoming acquainted with, for solving many problems in industry, are *fluid dynamics*, the modelling of fluid flows, important in a huge number of industries, *elasticity theory*, the modelling of solid structures, and *electromagnetic theory*, the modelling of electromagnetic fields, useful for modelling in telecommunications, including optical fibre communication.

Three mathematical areas we have not really touched on in this book, but which are of crucial importance to many, if not all, industries are: scientific computation, statistics, and operations research.

Scientific computation involves obtaining approximate numerical solutions to equations using a computer. An associated area of mathematics is ***numerical analysis*** which examines issues of efficiency, accuracy and stability of the algorithms used in scientific computation. Modern developments in software design and faster, parallel computers have meant that problems that were previously too difficult to solve numerically are now feasible. For anyone seriously contemplating a career in applying mathematics in industry a working knowledge of scientific computation is essential, as is some ability in computer programming. However, it is the authors' opinion that, sometimes, it is all too tempting to rush off and try to solve a problem numerically without trying to get analytic solutions (to possibly simpler models). We hope we have demonstrated in this book how analytic solutions can be of great value.

Statistics is another very important area. Statistical inference is the natural companion to experiment. It uses ideas from probability theory to test if results are significant or can be attributed to random variation.

Operations research encompasses a variety of techniques which are usually aimed at solving some sort of optimisation problem. Problems with constraints involve linear and nonlinear programming and many problems can be cast into this framework, such as scheduling, analysis of distribution networks. Typically these problems have a large number of variables.

A few other mathematical areas, important for many industrial problems, are described very briefly in the following.

- ***Probability theory:*** This includes the theory of probability distributions and the use of it for dealing with problems where random variation is significant. Techniques include Markov models (which use matrices), queueing models and stochastic differential equations (where the solutions are probability distributions).

- ***Integral equations:*** Some problems are more naturally formulated as integral equations (equations involving integrals) rather than differential equations. Special techniques are available for solving integral equations.

- ***Signal and image processing:*** This involves the analysis of signals (think of a signal as an time dependent quantity with some random component). An important tool for this is Fourier analysis, where a signal is decomposed into periodic components corresponding to sine and cosine functions and therefore extract patterns from noise. (Note that Fourier series were also used in a different context in Chapter 6.) Similar mathematics is used for manipulating inmages (such as satallite images).

- **Control theory:** Control theory is about systems with feedback where the aim is to drive the system to a certain state. In optimal control theory the aim is to optimise a quantity determined from a set of equations to determine some variable which characterises the feedback. A typical problem in control theory is to find fuel rate (the control variable) so that a car maintains a constant speed.

- **Discrete mathematics:** This includes various techniques for handling problems where the outputs are discrete quantities. These techniques include graph theory, set theory and number theory.

If, after reading this section, the reader feels a little daunted it is good to know that very few people in the world, if any, would claim expert knowledge in all these fields. We are only suggesting that a passing awareness of them is important to be able to recognise the possibilities that might occur when an appropriate problem from industry presents itself.

An excellent overview of many of the techniques mentioned briefly here can be found in Gershenfield (1999).

7.3 Mathematics in some other industries

Because of our focus on heat and mass transfer problems in this book it is natural that our case studies will have come predominantly from **manufacturing industries** (the main exception being the irrigation case study of Chapter 6 which was from **agriculture**).

Based upon the authors' experiences, some other important industries in which mathematics is extensively used include the following.

- **the food industry:** Heat and diffusion theory for problems with cooking and freezing, elasticity and fluid dynamics (e.g. making bread dough), dynamical systems for problems involving biochemical reactions involved with cooking, probability theory and population dynamics for deciding food safety issues, such as growth of bacteria.

- *finance:* This includes banking and insurance. Common problems involve quantifying risk and optimising investment portfolios. Mathematical techniques employed include probability theory, stochastic differential equations and operations research.

- *biomedical engineering:* Signal processing is used for artificial hearing devices. Fluid dynamics is used for the design of valves and artificial hearts.

- *telecommunications:* Queueing theory is used for the design of efficient communication networks and electromagnetic theory is used for the design of optical fibres and devices used to connect optical fibres.

- *mining:* This industry has problems involving blasting, scheduling and distribution. Mathematical techniques often used include statistics, dynamics, optimisation, elasticity.

- *transport:* This includes road, rail, air and sea. Problems include scheduling as well as solving structural problems (e.g. design of railway tracks).

- *environmental:* This is not really regarded as an industry in itself, however, many industry participants are seeing environmental issues to be of increasing importance for the solution of questions of sustainability, and there are many interesting mathematical problems that arise from environmental problems.

This chapter was designed to provide the reader with a taste of the possibilities of the use of mathematics in industry. The opportunities may only be limited by the reader's imagination!

References

Acheson, D. J. (1990). *Elementary Fluid Dynamics.* Oxford: Clarendon Press.

Andrews, J. G. and R. R. McLone (1976). *Mathematical Modelling.* UK: Butterworth.

Aziz, A. and T. Y. Na (1984). *Perturbation Methods in Heat Transfer.* Berlin: Springer-Verlag.

Barenblatt, G. I. (1987). *Dimensional Analysis.* Gordon and Breach.

Barton, N. D. (1985). Optimal control of a steel slab caster. In B. N. G and J. D. Gray (Eds.), *Mathematics in Industry Study Group Proceedings,* Australia, pp. 1–27. CSIRO Division of Mathematics and Statistics.

Batu, V. (1978). Steady infiltration from single and periodic strip sources. *Soil Science Society of America Journal 42,* 544–549.

Bear, J. (1972). *Dynamics of Fluids in Porous Media.* NY: Dover Publications.

Bedding, S. P. (1994). Concerning an integral arising in the study of laser drilling equations. *Int. J. Math. Educ. Sci. Tech. 25,* 669–672.

Birkhoff, G. (1950). *Hydrodynamics — A Study in Logic, Fact and Similitude* (1st ed.). Princeton, NJ: Princeton University Press.

Bluman, G. W. and S. Kumei (1989). *Symmetries and Differential Equations.* NY: Springer-Verlag.

Carslaw, H. S. and J. C. Jaeger (1959). *Conduction of Heat in Solids* (2nd ed.). UK: Oxford University Press.

Chang, Y.-Y. and M. Y. Corapcioglu (1997, May/June). Effects of roots on flow in unsaturated soils. *Journal of Irrigation and Drainage Engineering,* 202–209.

Crank, J. (1975). *The Mathematics of Diffusion.* (2nd ed.). London: Oxford University Press.

Dresner, L. (1983). *Similarity Solutions of Nonlinear Partial Differential Equations*. Boston: Pitman.

Drysdale, D. (1985). *An Introduction to Fire Dynamics*. NY: Wiley.

Edwards, D. and M. Hamson (1989). *Guide to Mathematical Modelling*. Macmillan.

Feynman, R. P., R. B. Leighton, and M. Sands (1977). *The Feynman Lectures in Physics* (7th ed.). USA: California Institute of Technology.

Fowkes, N. D. and J. J. Mahony (1994). *An Introduction to Mathematical Modelling*. UK: Wiley.

Fowler, A. C. (1997). *Mathematical Models in the Applied Sciences*. Cambridge University Press.

Frank-Kamenetskii, D. A. (1972). *Diffusion and Heat Transfer in Chemical Kinetics* (2nd ed.). NY: Plenum Press.

Fulford, G. R., P. Forrester, and A. Jones (1997). *Mathematical Modelling with Differential and Difference Equations*. Cambridge: Cambridge University Press.

Geiger, G. H. and D. R. Poirier (1980). *Transport Phenonema in Metallurgy*. (2nd ed.). Philippines: Addison-Wesley.

Gershenfield, N. (1999). *The Nature of Mathematical Modelling*. UK: Cambridge.

Gray, B. (1988). Effect of deposition of combustible matter onto electric power cables. In N. G. Barton (Ed.), *Proceedings of the 1988 Mathematics-in-Industry Study Group*, pp. 40–47. CSIRO, Australia.

Halliday, D. and R. Resnick (1974). *Fundamentals of Physics*. Canada: Wiley.

Hill, J. (1992). *Differential Equations and Group Methods for Scientists and Engineers*. Boca Raton : CRC Press.

Hill, J. M. (1987). *One-Dimensional Stefan Problems: An Introduction*. UK: Longman Scientific and Technical.

Hill, J. M. and J. N. Dewynne (1990). *Heat Conduction*. CRC Press.

Hill, J. M. and N. F. Smyth (1990). On the mathematical analysis of hot-spots arising from microwave heating. *Mathematics and Engineering in Industry 2*, 267–278.

Hillel, D. (Ed.) (1982). *Advances in Irrigation. Vol 1.*, NY. Academic Press.

Holman, J. P. (1992). *Heat Transfer* (5th ed.). Singapore: McGraw-Hill.

Holmes, M. H. (1995). *Introduction to Perturbation Methods*. NY: Springer-Verlag.

James, K. W., G. F. Thomson, and A. T. Hancock (1993). Evaluation of a survivor-06 water purification device. Technical Report MRL-TN-625, Materials Research Laboratory, DSTO.

Jones, J. C. (1993). *Combustion Science. Principles and Practice*. Brisbane:

Milennium Books.

Launder, B. E. and D. B. Spalding (1972). *Lectures in Mathematical Models of Turbulence*. London: Academic Press.

Lightfoot, E. N. (1974). *Transport Phenomena and Living Systems: Biomedical Aspects of Momentum and Mass Transport*. NY: Wiley.

Lin, C. C. and L. A. Segel (1974). *Mathematics Applied to Deterministic Problems in the Natural Sciences*. NY: Macmillan.

Logan, J. D. (1987). *Applied Mathematics: A Contemporary Approach*. NY: Wiley.

McElwain, D. L. S. and G. R. Fulford (1995). Moisture movement in bulk stockpiles. In J. Hewitt (Ed.), *Proceedings of the 1995 Mathematics-In-Industry Study Group*, pp. 48–65. University of South Australia.

McGowan, P. and M. McGuinness (1996). Modelling the cooking process of a single cerial grain. In J. Hewitt (Ed.), *Proceedings of the 1996 Mathematics-in-Industry Study Group.*, pp. 114–139. University of South Australia.

McNabb, A., C. Please, and D. L. S. McElwain (1999). Spontaneous combustion in coal pillars: Buoyancy and oxygen starvation. *Mathematical Engineering in Industry 7*, 283–300.

Mei, C. C. (1997). *Mathematical Analysis in Engineering*. UK: Cambridge.

Merten, U. (1966). *Desalination by Reverse Osmosis*. Cambridge University Press.

Nayfeh, A. H. (1981). *Introduction to Perturbation Techniques*. NY: Wiley.

Noble, B. (1967). *Applications of Undergraduate Mathematics in Engineering*. Mathematical Association of America.

Philip, J. R. (1970). Flow in porous media. *Annual Review of Fluid Mechanics 2*, 177–204.

Philip, J. R. (1988). Infiltration of water into soil. *Animal and Plant Sciences 1*, 231–235.

Probstein, R. F. (1989). *Physicochemical Hydrodynamics: An Introduction*. Boston: Butterworths.

Rowell, D. (1994). *Soil Science: Methods and Applications*. UK: Longman Group.

Shamsi, M. R. R. I. and S. P. Mehrotra (1993). A two-dimensional heat and fluid-flow model of a single-roll continuous-sheet casting process. *Metallurgical Transactions. 24B*, 521–535.

Sisson (1993). The self-heating of damp cellulosic materials. II. On the steady states of the spacially distributed case. *IMA J. Applied Mathematics 50*, 285–306.

Spiegel, M. R. (1968). *Mathematical Handbook*. Schaum Outline Series.

Stormont, J. C. and C. E. Morris (1997, Sep/Oct). Unsaturated drainage layers for diversion of infiltrating water. *Journal of Irrigation and Drainage Engineering*, 364–367.

Tayler, A. B. (1986). *Mathematical Models in Applied Mechanics*. Oxford: Clarendon Press.

Weber, R. O. and K. A. Renkema (1995). Spontaneous ignition in the presence of a power source. *Combustion Science and Technology 104*, 169–179.

Weil, C. (1988). Atmospheric dispersion — observations and models. In W. L. Steffen and O. T. Denmead (Eds.), *Flow and Transport in the Natural Environment: Advances and Applications*. Springer-Verlag.

Wigner, E. P. (1960). The unreasonable effectiveness of mathematics in the natural sciences. *Pure and Applied Mathematics 13*, 1–14.

Wilcox, D. C. (1994). *Turbulence Modelling for CFD* (Second ed.). Ca: DCW Industries.

Wilmott, P. (1998). *Derivatives. The Theory and Practice of Financial Engineering*. Chichester: Wiley.

Wilmott, P., S. Howison, and J. Dewynne (1995). *The Mathematics of Financial Derivatives: A Student Introduction*. UK: Cambridge.

Index